Learning Materials in Biosciences

Learning Materials in Biosciences textbooks compactly and concisely discuss a specific biological, bio-medical, biochemical, bioengineering or cell biologic topic. The textbooks in this series are based on lec-tures for upper-level undergraduates, master's and graduate students, presented and written by authoritative figures in the field at leading universities around the globe.

The titles are organized to guide the reader to a deeper understanding of the concepts covered.

Each textbook provides readers with fundamental insights into the subject and prepares them to independently pursue further thinking and research on the topic. Colored figures, step-by-step protocols and take-home messages offer an accessible approach to learning and understanding.

In addition to being designed to benefit students, Learning Materials textbooks represent a valuable tool for lecturers and teachers, helping them to prepare their own respective coursework.

More information about this series at http://www.springer.com/series/15430

Babak Arjmand
Moloud Payab
Parisa Goodarzi

Editors

Biomedical Product Development: Bench to Bedside

 Springer

Editors
Babak Arjmand
Cell Therapy and Regenerative Medicine
Research Center, Endocrinology and
Metabolism Molecular-Cellular Sciences
Institute
Tehran University of Medical Sciences
Tehran, Iran

Moloud Payab
Obesity and Eating Habits Research Center,
Endocrinology and Metabolism
Molecular-Cellular Sciences Institute
Tehran University of Medical Sciences
Tehran, Iran

Parisa Goodarzi
Brain and Spinal Cord Injury Research
Center, Neuroscience Institute
Tehran University of Medical Sciences
Tehran, Iran

ISSN 2509-6125   ISSN 2509-6133(electronic)
Learning Materials in Biosciences
ISBN 978-3-030-35625-5   ISBN 978-3-030-35626-2 (eBook)
https://doi.org/10.1007/978-3-030-35626-2

This Springer imprint is published by the registered company Springer Nature Switzerland AG
The registered company address is: Gewerbestrasse 11, 6330 Cham, Switzerland

For our parents
For *Rasta* and *Arvid*

Preface

As a multidisciplinary domain of science, biomedicine encompasses a wide spectrum of science from investigation to diagnosis and treatment of diseases. From the biomedicine point of view, different stages of a biomedical product development should be scientifically designed and performed. Actually, this broad domain monitors the whole processes of product development precisely from bench to bedside. Thanks to related international standards and frameworks, various concerns and obstacles have been diminished in the field of preclinical and clinical study design and also ethical considerations. Nowadays, experiments are designed and conducted according to GLP, GMP, and GCP standards. GLP has a pivotal role in preclinical studies, GMP principles have become critical requirements in manufacturing biomedical product for clinical uses, and, finally, clinical trials are performed based on unified GCP standards. Ultimately, the manufacturing of safe and efficient biomedical products can be achieved by adhering to current GxP standards and guidelines.

"Biomedical Products Development" is the title of an educational and practical course that has been established in Cell Therapy and Regenerative Medicine Research Center, affiliated to Endocrinology and Metabolism Molecular-Cellular Sciences Institute, Tehran University of Medical Sciences. In accordance with the mentioned course, this volume with the similar title has been published to elucidate our experiences in the field.

We would like to acknowledge Dr. Kursad Turksen, Consulting Editor of Learning Materials in Biosciences, for his kind support. Also, we would like to thank Amrei Strehl, Senior Editor of *Biochemistry and Cell Biology*, and Sivachandran Ravanan, Project Coordinator (Books) at Springer Nature, for their continuous efforts to get the volume to the print stage.

Babak Arjmand
Tehran, Iran

Moloud Payab
Tehran, Iran

Parisa Goodarzi
Tehran, Iran

Contents

Editors and Contributors

About the Editor

Babak Arjmand, M.D., Ph.D.

is the Head and Director of the Cell Therapy and Regenerative Medicine Research Center at the Endocrinology and Metabolism Molecular-Cellular Sciences Institute, Tehran University of Medical Sciences (Tehran, Iran). He graduated in Medicine from the Iran University of Medical Sciences (Tehran, Iran), and from 2011 to 2015 he completed his Ph.D. in Applied Cell Science at the Tehran University of Medical Sciences. His research interests focus on developing cell and gene-based clinical products through translational pathways, from basic investigations to clinical research in line with GLP, GMP, and GCP standards. Dr. Arjmand has trained a large number of biomedical students and young researchers. He is the author of more than 100 articles published in scientific journals and has also published various books, book chapters and conference papers. Dr. Arjmand is the member of different scientific committees and societies such as, Tissue Engineering and Regenerative Medicine International Society (TERMIS), Asia Pacific Association of Surgical Tissue Banks (APASTB), Iranian Tissue Engineering and Regenerative Medicine (ITERM), National Committee of Tissue, Cell and Gene Therapy at Iran Food and Drug Administration (Iran-FDA), Iranian Council of Stem Cell Technologies, and National Working Group for Providing National Guideline on Stem Cell therapy.

Moloud Payab, Ph.D.

obtained her B.Sc. and M.S. in Nutrition Sciences from Tehran Islamic Azad University in 2007 and Tehran University of Medical Sciences in 2011, respectively. Then she completed her Ph.D. at the Obesity and Eating Habits Research Center, Endocrinology and Metabolism Research Institute (EMRI), which is affiliated to Tehran University of Medical Sciences. She has worked as a researcher at EMRI since 2012. To date, her research interests have mostly focused on the field of clinical and population based studies from the endocrinology perspective, including but not limited to obesity. Further, she has been actively involved in designing, planning, implementing and monitoring several basic and clinical research projects. Her particular interest is cell therapy in the field of obesity using various cell-based products, and she closely collaborates with the Cell Therapy and Regenerative Medicine Research Center at EMRI.

Parisa Goodarzi, M.S.

obtained her B.Sc. in Nursing at Tehran University of Medical Sciences, Tehran, Iran in 1996. Then, she completed her M.S. in Geriatric Nursing at Iran University of Medical Sciences, Tehran, Iran in 2016. Since 2008, Parisa Goodarzi has worked as a researcher at Tehran University of Medical Sciences. She has extensive experience in the implementation of quality management systems (QMS) in the field of cell and tissue-based products and also product development pathways. She is also the author of more than 50 articles published in scientific journals, books and conference proceedings.

Contributors

Mina Abedi
Cell Therapy and Regenerative Medicine
Research Center, Endocrinology and
Metabolism Molecular-Cellular Sciences
Institute, Tehran University of Medical Sciences
Tehran, Iran
min.a1220@yahoo.com

Hossein Adibi
Diabetes Research Center, Endocrinology and
Metabolism Clinical Sciences Institute, Tehran
University of Medical Sciences
Tehran, Iran
adibi@tums.ac.ir

Hamid Reza Aghayan
Cell Therapy and Regenerative Medicine
Research Center, Endocrinology and
Metabolism Molecular-Cellular Sciences
Institute, Tehran University of Medical Sciences
Tehran, Iran
hr.aghayan@gmail.com

Sepideh Alavi-Moghadam
Cell Therapy and Regenerative Medicine
Research Center, Endocrinology and
Metabolism Molecular-Cellular Sciences
Institute, Tehran University of Medical Sciences
Tehran, Iran
sepidalavi@gmail.com

Maryam Arabi
Cell Therapy and Regenerative Medicine
Research Center, Endocrinology and
Metabolism Molecular-Cellular Sciences
Institute, Tehran University of Medical Sciences
Tehran, Iran

Metabolomics and Genomics Research Center,
Endocrinology and Metabolism
Molecular-Cellular Sciences Institute, Tehran
University of Medical Sciences
Tehran, Iran
Maryam.arabi98@gmail.com

Babak Arjmand
Cell Therapy and Regenerative Medicine
Research Center, Endocrinology and Metabo-
lism Molecular-Cellular Sciences Institute
Tehran University of Medical Sciences
Tehran, Iran
barjmand@sina.tums.ac.ir

Alireza Baradaran-Rafii
Ocular Tissue Engineering Research Center,
Shahid Beheshti University of Medical Sciences
Tehran, Iran

Ophthalmic Research Center, Shahid Beheshti
University of Medical Sciences
Tehran, Iran
alirbr@gmail.com

Najmeh Foroughi-Heravani
Cell Therapy and Regenerative Medicine
Research Center, Endocrinology and
Metabolism Molecular-Cellular Sciences
Institute, Tehran University of Medical Sciences
Tehran, Iran
najmeh_foroughi@yahoo.com

Firoozeh Ghaderi
Brain and Spinal Cord Injury Research Center,
Neuroscience Institute, Tehran University of
Medical Sciences
Tehran, Iran
firozehghaderi@yahoo.com

Mahdi Gholami
Department of Toxicology & Pharmacology,
Faculty of Pharmacy
Toxicology and Poisoning Research Center,
Tehran University of Medical Sciences
Tehran, Iran
m_gholami2068@yahoo.com

Kambiz Gilany
Department of Biomedical Sciences
University of Antwerpen
Antwerp, Belgium

Reproductive Immunology Research Center,
Avicenna Research Institute, ACECR
Tehran, Iran

Integrative Oncology Department
Breast Cancer Research Center, Motamed
Cancer Institute, ACECR
Tehran, Iran
kambiz.gilany@gmail.com

Parisa Goodarzi
Brain and Spinal Cord Injury Research Center,
Neuroscience Institute
Tehran University of Medical Sciences
Tehran, Iran
pr_goodarzi@yahoo.com

Mahdieh Hadavandkhani
Cell Therapy and Regenerative Medicine
Research Center, Endocrinology and
Metabolism Molecular-Cellular Sciences
Institute, Tehran University of Medical Sciences
Tehran, Iran
mahdiehhadavandkhani@gmail.com

Bagher Larijani
Endocrinology and Metabolism Research
Center, Endocrinology and Metabolism
Clinical Sciences Institute, Tehran
University of Medical Sciences
Tehran, Iran
emrc@tums.ac.ir

Maryam Moayerzadeh
Evidence Based Medicine Research Center,
Endocrinology and Metabolism
Clinical Sciences Institute, Tehran
University of Medical Sciences
Tehran, Iran
moayerzadeh@yahoo.com

Fereshteh Mohamadi-Jahani
Brain and Spinal Cord Injury Research Center,
Neuroscience Institute, Tehran University of
Medical Sciences
Tehran, Iran
f.mohamadijahani@gmail.com

Hamideh Moosapour
Evidence Based Medicine Research Center,
Endocrinology and Metabolism
Clinical Sciences Institute, Tehran
University of Medical Sciences
Tehran, Iran
dr.moosapour@gmail.com

Ensieh Nasli-Esfahani
Diabetes Research Center, Endocrinology
and Metabolism Clinical Sciences Institute,
Tehran University of Medical Sciences
Tehran, Iran
n.nasli@yahoo.com

Mehran Nematizadeh
Metabolomics and Genomics Research Center,
Endocrinology and Metabolism Molecular-
Cellular Sciences Institute, Tehran
University of Medical Sciences
Tehran, Iran
Mehran.nematizadeh@gmail.com

Moloud Payab
Obesity and Eating Habits Research Center,
Endocrinology and Metabolism Molecular-
Cellular Sciences Institute
Tehran University of Medical Sciences
Tehran, Iran
Moloudpayab@gmail.com

Fakher Rahim
Health Research Institute, Thalassemia and
Hemoglobinopathies Research Center, Ahvaz
Jundishapur University of Medical Sciences
Ahvaz, Iran
bioinfo2003@gmail.com

Ilia Rezazadeh
Cell Therapy and Regenerative Medicine
Research Center, Endocrinology and
Metabolism Molecular-Cellular Sciences
Institute, Tehran University of Medical Sciences
Tehran, Iran
ilia.rezazadeh@gmail.com

Masoumeh Sarvari
Cell Therapy and Regenerative Medicine
Research Center, Endocrinology and
Metabolism Molecular-Cellular Sciences
Institute, Tehran University of Medical Sciences
Tehran, Iran
maasoomehsarvari@yahoo.com

Motahareh Sheikh-Hosseini
Metabolomics and Genomics Research Center,
Endocrinology and Metabolism Molecular-
Cellular Sciences Institute, Tehran University of
Medical Sciences
Tehran, Iran
mt_sh_hosseini@yahoo.com

Akram Tayanloo-Beik
Cell Therapy and Regenerative Medicine
Research Center, Endocrinology and
Metabolism Molecular-Cellular Sciences
Institute, Tehran University of Medical Sciences
Tehran, Iran
a.tayanloo@gmail.com

Mehrnoosh Yarahmadi
Endocrinology and Metabolism Research
Center, Endocrinology and Metabolism
Clinical Sciences Institute, Tehran
University of Medical Sciences
Tehran, Iran
Mehrnoosh.y.a@gmail.com

Asal Zarvani
Evidence Based Medicine Research Center,
Endocrinology and Metabolism
Clinical Sciences Institute, Tehran
University of Medical Sciences
Tehran, Iran
asal_zvn1994@yahoo.com

List of Abbreviations

21-CFR	Code of Federal Regulation for Food and Drugs
ABSL	Animal facility biosafety level
AD	Alzheimer's disease
AIDS	Acquired immune deficiency syndrome
APA	American Psychological Association
ATMP	Advanced therapy medicinal products
BM-MSCs	Bone marrow mesenchymal stem cells
BMT	Bone marrow transplantation
BSC	Biological safety cabinet
BSC	Biological safety cabinets
BSL	Biosafety level
CAT	The Committee for Advanced Therapies
CBER	Center for Biologics Evaluation and Research
CBER	Center for Biologics Evaluation and Research
CCDS	Canine cognitive dysfunction syndrome
CDSCO	Central Drugs Standard Control Organization
CF	Cystic fibrosis
CFR	Codes of Federal Regulations
cGMP	Current good manufacturing practice
CGTPs	Cell and gene therapy products
CIOMS	Council for International Organization of Medical Sciences
CPCSEA	Control and supervision of experiments on animals
CRF	Case report form
CT	Computed tomography
CTOR	Safety of human cells, tissues, and organs for transplantation regulations
CTP	Cell therapy product
CTT	Cell- and tissue-based therapeutic
DI water	Deionized water

EAE	Experimental autoimmune encephalomyelitis
ECRIN	European Clinical Research Infrastructure Network
EDQM	The European Directorate for the Quality of Medicines & HealthCare
EMA	European Medicines Agency
EMA/EMEA	The European Medicines Agency
EMS	Equine metabolic syndrome
EPA	Environmental Protection Agency
ESCs	Embryonic stem cells
EU	European Union
EU	European Countries
F&DR	Food and Drug Regulation
FD&C Act	Federal Food, Drug, and Cosmetic Act
FDA	Food and Drug Administration
FDA	Federal Drug Administration
FDA	Food and Drug Administration
GCLP	Good clinical laboratory practices
GCP	Good clinical practice
GI	Gastrointestinal
GLP	Good laboratory practice
GMP	Good manufacturing practice
GMT	Good microbiological techniques
HCTPs	Human cells, tissues, and cellular and tissue-based products
HD	Huntington's disease
HEPA	High efficiency particulate air
HEPA	High-efficiency particulate air
HICs	High-income counties
HIV	Human immunodeficiency viruses
HSP	Human subject protection
HVAC	Heating ventilation and air conditioning
HVAC	Heating, ventilation, and air conditioning
IABS	International Alliance for Biological Standardization
IB	Investigator's brochure

ICCBBA	International Council for Commonality in Blood Banking Automation	NRAs	National Regulatory Authorities
ICH	International Conference on Harmonization	OCTGT	Office of Cellular, Tissue and Gene Therapies
IEC	Independent ethics committee	OGCP	Office of Good Clinical Practice
IND	Investigational New Drug	PD	Parkinson's disease
IND	Investigational new drug	PET	Positron emission tomography scans
iPSCs	Induced pluripotent stem cells		
IRB	Institutional Review Board	PHS Act	Public Health Service Act
ISA	International Federation of the National Standardizing Associations	PMDA	Pharmaceuticals and Medical Devices Agency
ISBT	International Society of Blood Transfusion	QA	Quality assurance
ISO	International Organization for Standardization	QC	Quality control
		QMS	Quality management system
ISO	International Standards Organization	QSR	Quality Systems Regulations
ISPE	International Society for Pharmaceutical Engineering	RCT	Randomized clinical trial
		REC	Research ethics committee
K-FDA	South Korean Food and Drug Administration	SOP	Standard operating procedure
		SOPs	Standard operating procedure
LMICs	Low- and middle-income countries	TGA	Therapeutic Goods Administration
LTCs	The MOH CTT licensing terms and conditions		
		U.S. FS 209	United States Federal Standard 209
MCP	Minimum consensus package	ULPA	Ultra low particulate air
MD	Medical device	UNSCC	United Nations Standards Coordinating Committee
MDR	Medical devices regulations		
MHRA	Medicines and Healthcare Products Regulatory Agency	VDEGS	Van den Ende-Gupta syndrome
MOH	Ministry of Health	WHO	World Health Organization
MRI	Magnetic resonance imaging	WMA	World Medical Association
MSCs	Mesenchymal stem cells	WWII	World War II

An Introduction to Biomedical Product Development

Babak Arjmand (iD), *Moloud Payab* (iD), *and Parisa Goodarzi*

References – 3

© Springer Nature Switzerland AG 2020
B. Arjmand et al. (eds.), *Biomedical Product Development: Bench to Bedside*,
Learning Materials in Biosciences, https://doi.org/10.1007/978-3-030-35626-2_1

1

» A man will be imprisoned in a room with a door that's unlocked and opens inwards; as long as it does not occur to him to pull rather than push it. – Ludwig Wittgenstein

When the fundamental structures of science were created, an important stage in research began to rise. The concept behind this stage was remembering to think critically. This history got separated into three parts: (1) the early phase based on ritual and magic, (2) the rational phase based on the creative imagination, and (3) the modern phase based on experimental designs and laboratory investigations. In this context, the Chinese began to study their traditional medicine in the first century BC. They wrote "the yellow emperor's book of medicine," the traditional Chinese medicine book. This was a good initiation to research biology and medical sciences [1].

After a few centuries, the path of science with all its ups and downs has made many changes in the process of studies and the tests that take place today. Following identifying various microorganisms and their pathogenicity, the need to deal with them was felt.

Thus, a more accurate diagnosis before treatment was needed for developing varies diagnostic and therapeutic approaches, and finally, antibiotics were born. Further, the development of basic sciences and the connections between its different branches lead to a greater expansion of knowledge boundaries. The emersion of interdisciplinary science today is attributed to these developments through different periods of time.

As one of the main branches of science, biomedical one includes a set of applied sciences that integrates technology and various elements of knowledge. It contains fundamental fields such as anatomy, physiology, biology, biochemistry, biophysics, immunology, mathematics, statistics, bioinformatics, etc. Biomedical sciences first got defined by the UK Quality Assurance Society [2]. This approach includes scientific disciplines focusing on biology, public health, and human diseases. It can also get more specialized in topics such as pharmacology, human physiology, and human nutrition [2].

On the other hand, biomedical sciences encompass a wide range of research and scientific activities that have significant economic and scientific impacts over the hospitals and laboratory sciences. In summary, the vital role of biomedical science and its products in the development of infrastructure and economics in research centers, institutes, and companies has been depicted clearly over the years.

Today, many companies including pharmaceutical corporations are competing for hundreds of awards across the globe. In order to ensure that users are receiving the highest quality products, quality management system (QMS) which is one of the most important goals of each company or institute must be established to help with achieving quality aims [3]. Fortunately, the research in the biomedical area has been greatly developed in recent years. Thus, it has led to having higher quality biomedical products. To achieve and maintain this situation, several standards, rules, and policies have emerged in order to promise the uppermost quality.

This book tries to describe the life cycle of biomedical products from the idea to the application. For instance, some of the most important issues about the essential of QMS and its requirements in biomedical sciences are discussed with the design of the instruments in this field (e.g., the importance of cleanrooms in biomedical products).

Further, it talks about the design of experimental studies, performance, and monitoring of clinical trials, standards, and frameworks in the field and the principles of good laboratory practice (GLP), good clinical practice (GCP), and good manufacturing practice (GMP).

According to the GLP principles, experimental studies as the fundamental stage should be assured for quality and integrity. For this purpose, it is important to design and conduct standard preclinical studies. Furthermore, a set of regulations and principles of GMP are required to ensure the quality of manufactured products. Additionally, clinical research is a pivotal step in the approval process of developing a biomedical product. Since clinical trials are involved with human subjects, it will be raised several ethical concerns and considerations from design to audit and approve the study and its results. Accordingly, GCP principles provide unified standards to simplify clinical data acceptance by relevant authorities.

Regardless of the reader having professional experience in biomedical sciences and related areas to developing standard research, this book will be a valuable asset to everyone and will cover all steps of manufacturing a biomedical product. Also, it will broaden the reader's horizon and their understanding in the field to address the importance of biomedical principles in product development from bench to bedside.

References

1. Curran J. The yellow emperor's classic of internal medicine. BMJ. 2008;336(7647):777.
2. Setting and maintaining academic standards. The Quality Assurance Agency for Higher Education. 2018. https://qaa.ac.uk. Accessed April 10, 2019.
3. Special issue: abstracts from the 2008 ISSOL meeting. Orig Life Evol Biosph. 2009;39(3):179–392.

Basic Essentials and Applications of Quality Management System (QMS) in Biomedical Sciences

Babak Arjmand (iD), *Motahareh Sheikh-Hosseini, Fakher Rahim, Hossein Adibi, Alireza Baradaran-Rafii, and Bagher Larijani*

© Springer Nature Switzerland AG 2020
B. Arjmand et al. (eds.), *Biomedical Product Development: Bench to Bedside*,
Learning Materials in Biosciences, https://doi.org/10.1007/978-3-030-35626-2_2

2.1 What You Will Learn in This Chapter

- The importance of quality management system (QMS) in biomedical sciences
- The implementation of the international quality standards for laboratories
- The importance of implementing safety management programs
- The importance of trained and qualified personnel
- The implementation of quality control processes and quality assurance
- The importance of monitoring, analyzing, and managing of the QMS
- The necessity of continuous improvement
- The necessity of establishing a program to address user requirements and satisfaction
- The functions and requirements of the organizational process, management, and structure
- The necessity of the documentation including keeping records and documents
- The challenges in QMS and applying new techniques in the near future

2.2 Rationale and Importance

Although the essentials of the quality management system (QMS) were introduced in the 1920s, its implementation in the laboratory was started in the 1940s. Quality in the laboratory means to have a traceable and accurate test results. Accordingly, any level of inaccuracy should be avoided. On the other hand, in the field of biomedical product manufacturing, achieving a safe and efficient product needs implementing all of the aspects of the quality in the manufacturing process [1, 2].

Depending on the type of project, in all fields including diagnostic laboratory management, biomedical product management, scientific research, and clinical trials, the implementation of guidelines on good laboratory practice (GLP), good manufacturing practice (GMP), and good clinical practice (GCP) should be considered for achieving acceptable standards in the quality management process (Fig. 2.1) [3, 4]. GLP refers to the set of methods which regulates the processes and activities in the laboratory to ensure the reliability and safety of laboratory data [5]. On the other hand, GMP is the set of regulations that monitor the manufacturing processes to improve the quality, purity, and safety of products [6]. Finally, promoting the safety, reliability, and integrity of data reporting and methodology associated with clinical trials is implemented by GCP [7].

2.3 A Snapshot of the Quality Management System

The structure of the QMS is made of some essentials that should work together and be managed properly. QMS can control and direct an organization in accordance with quality features. In this organization, personnel are responsible for achieving the highest quality and continuous improvement in every organizational position [8, 9].

Fig. 2.1 Commitment to GLP, GMP, and GCP in biomedical projects can improve the quality of data and products as the final conclusion of QMS implementation

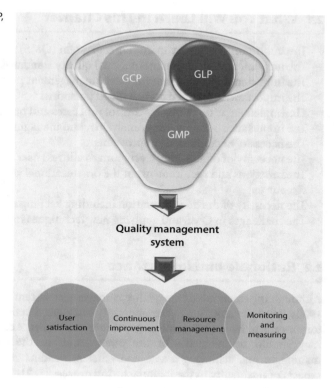

2.3.1 The Definition of the Quality Management System

QMS is a set of procedures, activities, resources, policies, documented information, and organized processes that help organizations to meet quality improvement, consistency, and good customer services. QMS regulates, coordinates, and directs the activities of the organization in order to meet regulatory and user needs, to improve its efficiency and effectiveness continuously [8, 9].

2.3.2 Quality Management System Principles in Biomedical Sciences

The aim of QMS implementation in biomedical sciences is to improve bioproduct characteristics, productivity, efficiency, public awareness, working atmosphere, continuous improvement, as well as lowering costs. In summary, commitment to the QMS principles can improve organizational performance. Some of the major principles of QMS are as follows [9, 10]:
- User focus
- Leadership
- People involvement
- Process approach
- System approach to management

- Continuous improvement
- Factual decision-making
- Mutually beneficial supplier relationships

2.4 An Introduction to International Organization for Standardization

In 1946, a meeting has been held between the International Federation of the National Standardizing Associations (ISA) and the United Nations Standards Coordinating Committee (UNSCC) about setting the International Organization for Standardization (ISO) to make worldwide series of international standards. These standards are known as ISO standards [11]. Then, in 1947, different countries decided to establish ISOs as nongovernmental organizations and set commercial and industrial standards to improve quality features of products and services and also to provide appropriate solutions for global challenges [9, 11]. In summary, ISO is a nongovernmental organization that develops a series of international standards and works with different institutes in around 150 countries [11].

2.5 International Quality Standards for Laboratories

Assessing performances according to quality standards has a crucial role in implementing QMS. Standards should be applied to meet the regulatory requirements and user needs for enhancing the safety, consistency, and also monitoring of laboratory performances [11]. Therefore, a group of standards has been published to ensure that the production of biomedical products is fit for their goals. Accordingly, ISOs have some different series based on their aims and scopes [12]:

- ISO 9000:2015 focuses on QMS fundamentals and vocabulary and also provides a guidance for different kinds of organizations to promote quality management and productivity as well as address user requirements [12].
- ISO 9001:2015 specifies QMS requirements as well as providing the standards for achieving user satisfaction [12, 13].
- ISO 15189:2012 defines requirements of competency and quality in medical laboratories [14].
- ISO 17025:2017 focuses on the general requirements of calibration laboratories and competence of testing [15].
- ISO 13485:2016 identifies the QMS requirements to provide medical devices [11].
- ISO 14644:2019 focuses on cleanroom classification and associated controlled environments [16].

2.6 Implementing of Safety Management Programs

Safety management programs as one of the essential parts of the QMS contain safety of personnel, facilities, equipment, and manufactured biomedical products. Safety equipment and facilities are used for achieving quality and safety of products, personnel, and patients and also for preventing the chemical and biological hazards [9]. In this context, relevant policies and standard procedures must be established. Accordingly, personnel should be aware of the potential hazards and be trained for applying safety practices [2]. Additionally, laboratory safety program can optimize and improve health status of the patients and also working atmosphere [2]. On the other hand, the reputation of the institutes will be improved, and the income will be increased using QMS. Finally, QMS can help the organization to reduce its undue costs and negative effects [17].

2.7 Personnel Training Program

Personnel has a crucial role in every organization. Trained, qualified, and responsible personnel help organizations to achieve their goals such as high quality of biomedical products and services. As a result, health status of the patients will be improved [18]. Therefore, the accurate data and high-quality services depend on the competent personnel who have the ability to perform different tasks and procedures properly. Accordingly, training process should focus on personnel skills knowledge and behavior [18]. First of all, competency assessment such as evaluating education of all employees and also their knowledge and problem-solving skills must be assessed before starting training programs. In addition, it is necessary to write a policy for competency assessment. Further, all the personnel should know about the policy of the organization. Finally, all the procedures in the organizations should be documented [19].

2.8 Quality Control Processes

Quality control processes use monitoring to ensure the accuracy and reliability of process and include the quality control (QC) and sample management process [20].
- *Quality Control*
- QC is one of the main components of the process control. It can monitor, examine, and detect failures that may occur throughout the processes [20].
- *Sample Management Process*
- Accuracy and reliability of quality data and test results depend on the suitable management of samples. Subsequently, the quality of the results has a significant effect on clinical decisions, patient care, and good treatment [9]. Therefore, written protocols for sample management are necessary considering some important points such as [2] the following:
  - Sample collection
  - Sample preservation
  - Sample labeling
  - Sample storage and retention
  - Sample transport
  - Sample disposal

2.9 Quality Assurance

Quality assurance (QA) is a set of systematically structured plans and activities to ensure the quality of products and services. Implementing QA can reduce problems and defects in manufactured products and laboratory results [21]. Personnel should be aware of all the QA procedures and be committed to all relevant requirements and standards [9].

2.10 Monitoring Based on the Quality Management System

Monitoring, analyzing, and managing are pivotal in evaluating quality of an organization based on QMS features [2]. Further, monitoring in QMS should be performed in accordance with established standards such as ISOs. Therefore, each process should be monitored using some appropriate methods to increase its effectiveness [9].

2.11 Continuous Improvement

Since continuous improvement is critical in QMS, monitoring and evaluating the effectiveness and performances of all activities in organizations step by step are necessary for detecting weaknesses and errors and correcting them [11]. Continuous improvement mainly focuses on improving and increasing activities to fulfill quality requirements. Accordingly, all the processes should be better continuously [9].

2.11.1 Implementing Tools for Process Improvement

There are a lot of practical techniques that are used in process improvement to evaluate, control, and detect the problems and defects in the processes, using external and internal audits [22].

2.11.1.1 External and Internal Audit

According to ISOs that focus on assessment requirements, the organization performance is evaluated. Therefore, suitable and practical tools should be applied as audits. Audits have a regulated program which detects and evaluates problems throughout the whole organization using two approaches [23]: (1) internal audit and (2) external audit. Internal audit is a type of assessment conducted by organization with a group of personnel. External audit is conducted by groups and agencies from outside the organization for getting accreditation, approval, licensure, and certification [9, 11].

2.12 Good Customer Services

The main goal of the QMS is to meet the user requirements. Institutes should be aware of the user needs and try to meet them by focusing on the ISO series to improve the quality of products and achieve the satisfaction of the users (e.g., patients, communities, physicians, and public health agencies). It is essential to establish a program to address user requirements and satisfaction [24]. In accordance with, all personnel should be committed to the system, using proper monitoring tools to collect necessary information. Therefore, they need to be trained for using these tools properly and also using technologies such as computer and Internet to document and save time [25].

2.13 Organizational Requirements and Functions of the Quality Management System

Organization includes management and organizational structure. Management should direct the organization by a visible support to show the importance of personnel efforts [26]. Its policies should be written according to the mission and factual-based decision-making. Moreover, the design of the organizational structure should meet all the quality requirements and be clearly defined. It is also important to ensure that all functions in the organization are performed properly in accordance with quality features [19].

2.14 Documentation

Documentation is a tool used for writing information about procedures, processes, and policies. Perfect documentation system needs clear, accurate, feasible, and practical documents and manuals which can be used by personnel easily [11]. By keeping records and documents, the information whenever is necessary will be available. Defining processes should be recorded to prove that they are being followed. Documentation has a pivotal role in objective evidences which can support personnel via manufacturing, design, and development of biomedical product to meet the requirements [9]. All the laboratories and manufacturing facilities should have clear standard operating procedures (SOPs) which are a type of document and instruction that should be followed by personnel. It is important to change and update the documents regularly [27]. Accordingly, policy is a type of fundamental document that includes main objectives which are defined in organizations and approved by the management system. It is necessary that each organization writes their specific policies according to national policies [19].

2.15 Challenges and Future Perspectives

Establishing a perfect QMS in an organization has some serious limitations. For instance, trained, expert, and committed personnel, implementing proper and up-to-date data protocols, easy access to suitable information, doing everything at the first time, avoiding time and budget waste, etc., are considerable challenges (◻ Table 2.1) [28, 29].

Some new techniques have been developed to improve institutes and organizations according to QMS specifications continuously. Accordingly, using such new methods and tools will not be negligible in the near future. In this context, lean process [30, 31] and Six Sigma [32, 33] are the most commonly used tools. Both of them are used for optimizing all activities of the organization to save time and money and also to improve organizational performances by controlling, analyzing, and measuring as well as to reduce the mistakes and problems in the processes [30, 32, 33].

■ Table 2.1 Some important challenges of the QMS implementation and possible solutions

Challenges	Solutions
Not enough satisfying users	Focusing on the user requirements
Too much unnecessary documentation	Implementing useful documentation management procedures
Too much focus on theory which is not practical	Focusing on practical theories and putting them into practice
Wasting too much time and resources on unnecessary details	Using details which are based on your activities
Too inflexible program	Implementing changes that improve the process
Not enough support and communication	Defining leadership and personnel responsibility clearly

---

Take-Home Messages

- QMS is a set of procedures, activities, resources, and organized processes to achieve quality specifications and fulfill user requirements.
- Implementing guidelines on GLP, GMP, and GCP in institutes can increase quality of data and products in accordance with relevant standards.
- ISO series intends to promote the quality, efficiency, and safety of biomedical products and technologies for users.
- Safety management programs can improve the safety of manufactured products, personnel, and patients.
- Training program can develop the skills, knowledge, and behaviors of the personnel.
- Quality control processes can control activities and monitoring processes to ensure continuous improvement.
- Continuous improvement is a permanent objective of an organization.
- Internal and external audits are the important tools of continuous improvement.
- The organizational requirements include proper management and also clear organizational structure to meet all the quality requirements.
- Institutes and organizations should be aware of the user needs and try to meet them by focusing on the ISOs.
- Documentation is a set of documents that includes writing procedures, policies, and processes which record all the necessary information and should be clear and easily accessible.
- Establishing a perfect QMS includes some challenges.

2

References

1. Weckenmann A, Akkasoglu G, Werner T. Quality management–history and trends. TQM J. 2015;27(3):281–93.
2. Allen LC. Role of a quality management system in improving patient safety—laboratory aspects. Clin Biochem. 2013;46(13–14):1187–93.
3. Geijo F. Quality management in analytical R&D in the pharmaceutical industry: building quality from GLP. Accred Qual Assur. 2000;5(1):16–20.
4. Suzuki-Nishimura T. Clinical trials and good clinical practice. J Health Sci. 2010;56(3):231–8.
5. Kendall G, Bai R, Błazewicz J, De Causmaecker P, Gendreau M, John R, et al. Good laboratory practice for optimization research. J Oper Res Soc. 2016;67(4):676–89.
6. McGowan NW, Campbell JD, Mountford JC. Good manufacturing practice (GMP) translation of advanced cellular therapeutics: lessons for the manufacture of erythrocytes as medicinal products . Erythropoiesis: Springer. Methods Mol Biol. 2018;1698:285–92.
7. Gumba H, Waichungo J, Lowe B, Mwanzu A. Implementing a quality management system using good clinical laboratory practice guidelines at KEMRI-CMR to support medical research. Version 2 Wellcome Open Res. 2018;3:137.
8. Kubono K. Quality management system in the medical laboratory--ISO15189 and laboratory accreditation. Rinsho Byori. 2004;52(3):274–8.
9. Organization WH. Laboratory quality management system: handbook. Geneva: World Health Organization; 2011.
10. Nanda V. Quality management system handbook for product development companies: CRC Press. London. 2016.
11. Organization WH. Laboratory quality standards and their implementation. WHO Regional Office for the Western Pacific: Manila; 2011.
12. Iso.Org [homepage on the internet]. Switzerland: International Organization for Standardization, [updated 2019 June, cited 2019 June 15]. Available from: https://www.iso.org.
13. Ingason HT. Best project management practices in the implementation of an ISO 9001 quality management system. Procedia Soc Behav Sci. 2015;194:192–200.
14. Theodorou D, Giannelos P. Medical laboratory quality systems - a management review. Int J Health Care Qual Assur. 2015;28(3):267–73.
15. Grochau IH, ten Caten CS. A process approach to ISO/IEC 17025 in the implementation of a quality management system in testing laboratories. Accred Qual Assur. 2012;17(5):519–27.
16. Shan K, Wang S. Energy efficient design and control of cleanroom environment control systems in subtropical regions–a comparative analysis and on-site validation. Appl Energy. 2017;204:582–95.
17. Ndihokubwayo J-B, Maruta T, Ndlovu N, Moyo S, Yahaya AA, Coulibaly SO, et al. Implementation of the World Health Organization Regional Office for Africa stepwise laboratory quality improvement process towards accreditation. Afr J Lab Med. 2016;5(1):1–8.
18. Berte LM. Laboratory quality management: a roadmap. Clin Lab Med. 2007;27(4):771–90.
19. Wadhwa V, Rai S, Thukral T, Chopra M. Laboratory quality management system: road to accreditation and beyond. Indian J Med Microbiol. 2012;30(2):131.
20. Carey RB, Bhattacharyya S, Kehl SC, Matukas LM, Pentella MA, Salfinger M, et al. Implementing a Quality Management System in the Medical Microbiology Laboratory. Clin Microbiol Rev. 2018;31(3).
21. Taylor JK. Quality assurance of chemical measurements, New York. 2018.
22. Picarillo AP. Introduction to quality improvement tools for the clinician. J Perinatol. 2018;38(7):929–35.
23. Beckmerhagen I, Berg H, Karapetrovic S, Willborn W. On the effectiveness of quality management system audits. TQM Mag. 2004;16(1):14–25.
24. Bergman B, Klefsjö B. Quality from customer needs to customer satisfaction: Studentlitteratur AB; 2010.
25. Iacob E. Experience of accreditation in a surface science laboratory. Accred Qual Assur. 2016;21(1):9–17.
26. Zaretzky AN. Quality management systems from the perspective of organization of complex systems. Math Comput Model. 2008;48(7–8):1170–7.
27. Barbé B, Verdonck K, Mukendi D, Lejon V, Kalo J-RL, Alirol E, et al. The art of writing and implementing standard operating procedures (SOPs) for laboratories in low-resource settings: review of guidelines and best practices. PLoS Negl Trop Dis. 2016;10(11):e0005053.

28. Sacchini F, Freeman KP. Quality documentation challenges for veterinary clinical pathology laboratories. J Vet Diagn Investig. 2008;20(3):266–73.
29. Williams R, Van Der Wiele T, Van Iwaarden J, Bertsch B, Dale B. Quality management: the new challenges. Total Qual Manag Bus Excell. 2006;17(10):1273–80.
30. Clark DM, Silvester K, Knowles S. Lean management systems: creating a culture of continuous quality improvement. J Clin Pathol. 2013;66(8):638–43.
31. Hussain A, Stewart LM, Rivers PA, Munchus G. Managerial process improvement: a lean approach to eliminating medication delivery. Int J Health Care Qual Assur. 2015;28(1):55–63.
32. Elbireer A, Le Chasseur J, Jackson B. Improving laboratory data entry quality using Six Sigma. Int J Health Care Qual Assur. 2013;26(6):496–509.
33. Westgard JO, Westgard SA. Six Sigma Quality Management System and design of risk-based statistical quality control. Clin Lab Med. 2017;37(1):85–96.

Further Reading

Books

Organization WH. Laboratory quality management system: handbook. Geneva: World Health Organization; 2011.
Organization WH. Laboratory quality standards and their implementation. Manila: WHO Regional Office for the Western Pacific; 2011.
Nanda V. Quality management system handbook for product development companies: CRC press; 2016.

Online Resources

https://europepmc.org/abstract/med/15137330, Quality management system in the medical laboratory--ISO15189 and laboratory accreditation.
https://www.ncbi.nlm.nih.gov/pubmed/25860923, Medical laboratory quality systems - a management review. https://doi.org/10.1108/IJHCQA-04-2014-0039.
https://www.ncbi.nlm.nih.gov/pubmed/17950897, Laboratory quality management: a roadmap. https://doi.org/10.1016/j.cll.2007.07.008.

Principles of Good Laboratory Practice (GLP)

Motahareh Sheikh-Hosseini, Parisa Goodarzi, Hamid Reza Aghayan, Kambiz Gilany, Firoozeh Ghaderi, Mahdi Gholami, and Babak Arjmand (iD)

© Springer Nature Switzerland AG 2020
B. Arjmand et al. (eds.), *Biomedical Product Development: Bench to Bedside*,
Learning Materials in Biosciences, https://doi.org/10.1007/978-3-030-35626-2_3

3.1 What You Will Learn in This Chapter

- The importance of understanding necessary information about good laboratory practice (GLP) as well as implementing relevant standard methods
- The importance of maintaining biosafety in the laboratory and applying correct assessment of the potential risks in laboratory
- The necessity of providing safety of individuals, environment, and chemicals in nonclinical laboratories
- The necessity of conducting GLP guidelines under carefully controlled situations and also traceable, reproducible, and reliable data set
- The importance of implementing standardized and validated procedures to support human clinical trials
- The necessity of setting GLP guidelines by different regulatory agencies to consider organizational structure, documentation (record keeping), quality assurance, personnel training practices, responsibilities, good microbiological techniques (GMT), safety equipment, standard operating procedures (SOPs), records, and sample retention

3.2 Rationale and Importance

In 1972, there were some cases of poor laboratory practice in the United States. Therefore, the FDA decided to do some in-depth researches of 40 toxicology laboratories and increase the number of laws for chemical and pharmaceutical products [1]. They discovered a lot of fraudulent activities and poor laboratory practices such as uncalibrated equipment, wrong measurements, inaccurate data, and inadequate test systems. Accordingly, by setting GLP guidelines, final data can demonstrate a true reflection of results gained during the study [2]. One of the important goals of GLP is promoting safety, quality, consistency, and reliability of products, data, and services in the process of laboratory testing, to improve human health and environmental risk management. Moreover, it is important to know how scientists use quality setting to improve biological products and data. Finally, research authorities should prove that there are no changes in the quality of data [1, 3].

3.3 Good Laboratory Practice (GLP)

3.3.1 An Overview of General Rules

GLP is a set of techniques that provide safety of personnel, laboratory, and environment and also pave the way for better laboratory practices by eliminating poor practices. Safety assessment based on GLP guidelines is a key step before starting clinical trials [4]. Therefore, GLP guidelines have conducted to guarantee biosafety in laboratories [5, 6]. Accordingly, this chapter will describe biosafety aspects of GLP focusing on risk management approaches that should consider risk groups [7]. In accordance with risk groups, each country should design a national classification of microorganisms based on their possibility of increasing harm in humans, animals, and environments according to the following issues [6]:

- Pathogenicity of the microorganism
- Transmission modes and host range of the organism
- Access to effective prevention plans which usually include prophylaxis by immunization, antisera administration, and sanitary measures
- Access to effective treatment measures including antimicrobials, antivirals, and chemotherapeutic agents, passive immunization, and post-exposure vaccination [6, 8]

There are four risk groups based on the classification of World Health Organization (WHO). Infectious microorganisms are categorized in risk groups based on their relative risks. Risk group 1 includes organisms which do not cause disease in humans or animals, risk group 2 pathogens cause diseases such as infections in humans or animals, risk group 3 agents cause serious diseases, and finally risk group 4 organisms cause lethal diseases in humans [9]. The classification of risk groups is used for laboratory work only. In contrast to risk groups, biosafety levels prescribe practices, facility requirements, engineering controls, and personal protective equipment (PPE) for working with infectious microorganisms. Similar to risk groups, there are four biosafety levels which are level 1, 2, 3, and 4 (◘ Table 3.1) [4, 6].

◘ **Table 3.1** The association between biosafety levels, laboratory types, arrangements, and equipment [4, 6, 7]

Risk groups	Biosafety level	Laboratory type	Classification of infectious microorganisms	Training requirements and equipment	Treatment procedures
1	Biosafety level 1	Basic laboratory 1	Microorganisms do not cause disease in humans or animals	Pass health test steps, GMT*	–
2	Biosafety level 2	Basic laboratory 2	Microorganisms cause disease such as infections in humans or animals	GMT, protective clothing, biohazard sign	Disease is preventable
3	Biosafety level 3	Containment laboratory	High concentration of group 2 microorganisms and cause serious diseases	As level 2 plus special clothing, controlled access, directional airflow	Treatments and vaccines
4	Biosafety level 4	Maximum containment laboratory	Microorganisms cause deadly disease in humans, and they can transfer from one person to another	As level 3 plus airlock entry, shower exit, special waste disposal	No treatments or vaccines

GMT good microbiological techniques

3.3.2 **Good Laboratory Practice Training**

Trained and well-organized personnel play a key role in the successful performance of procedures in a laboratory. On the other hand, it is necessary to have systems that assess persistent competency and trained personnel to convince suitable responsibility and communication during study conduct. Therefore, all personnel should receive direct and accurate training information to complete and perform their tasks [10]. After initial staff training, qualification assessments should be performed and recorded for all components of the training and practical responsibilities. As a result, the clinical education program should meet all the personnel needs and be documented and available to all laboratory staff [10, 11].

3.4 **Biosafety Concepts in the Laboratory**

Laboratory biosafety refers to protective measures, containment principles, and relevant technologies to prevent accidental exposure to pathogens and toxins. The association between four biosafety levels and infectious microorganisms based on their potential risks are demonstrated in ◘ Fig. 3.1. Effective implementation of biosafety in a laboratory is the base of laboratory biosecurity [7]. Accordingly, laboratory biosecurity refers to personal and organizational security measures that prevent misuse, loss, diversion, or deliberate release of pathogens and toxins [12]. On the other hand, risk assessment is an essential part of a biosafety program, which collects information about the type of available organisms, their physical location, and also identification of staff who are responsible for the

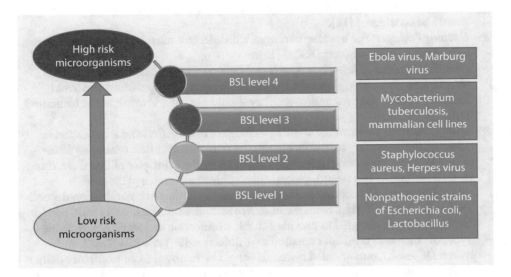

◘ **Fig. 3.1** Biosafety levels (BSL). Microorganisms classified by four risk groups and accordingly biosafety levels are divided into four classes. Biosafety level 1 refers to low-risk microorganisms such as Nonpathogenic strains of *Escherichia coli* and *Lactobacillus*, while biosafety level 4 refers to high-risk microorganisms such as *Ebola* and *Marburg* viruses. Additionally, biosafety levels 2 and 3 are located between 1 and 4 [7, 9]

maintenance of organisms [13]. Moreover, professional and ethical competency of all employees faced with dangerous pathogens that are permanently accessed to sensitive materials has a central role in laboratory biosafety programs. As a result, assessing competency of employees, training specific security issues, and complying accurate protection approach against pathogens are rational tools for promoting biosecurity in the laboratories [12]. Therefore, it is necessary to determine the biosafety level, type of microorganisms, available facilities, arrangements, skills, and procedures for performing a safe performance in laboratories [4, 14].

3.5 Good Microbiological Techniques

The aim of good microbiological techniques (GMT) is improving public health and also lifestyle. Each laboratory must take safety measures to eliminate or reduce potential hazards. GMT plays a pivotal role in laboratory safety and is based on the prevention of contamination [15].

Because of improper collecting, transferring, and handling of samples in laboratory, humans will be prone to risk of infections [16]. To prevent contamination in laboratories, some procedures should be considered as follows:

— *Transferring of Infectious Substances*: Transportation of infectious substances and materials should be done in accordance with national and international standards and rules. These rules describe how to properly use packaging materials and other transportation requirements to reduce potential damages and possible infections. For example, using a triple packaging system is essential for transporting potentially infectious substances [17].

— *Specimen Receipt*: A particular room for receiving a large number of specimens should be considered [18].

— *Opening Packages*: For opening packages, a disinfectant must be available and opened in biological safety cabinets (BSCs) [18].

— *Avoiding Ingestion of Infectious Agents and Contact with Skin and Eyes*: Personnel should wear disposable gloves during microbiological manipulations and avoid touching mouth and eyes; moreover, they should not eat and drink in the laboratory [16].

— *Disinfection and Sterilization*: Basic knowledge of disinfection and sterilization is essential for laboratory safety. As highly contaminated materials cannot be disinfected or sterilized quickly, it is important to know the principles of initial cleansing as disinfecting. In this regard, general rules apply to all known pathogens. Accordingly, specific conditions should be used to eliminate contamination depending on the type of testing, nature, and source of contaminations. As a result, preclearing materials are essential to provide suitable disinfection and sterilization. Some types of chemical germicides are used as disinfectant (◼ Table 3.2) [7, 17].

— *Waste Disposal*: Contaminated materials should be removed from laboratory daily. Moreover, most instruments, laboratory clothes, and glassware should be reused or recycled. It is important that all infectious materials are decontaminated, incinerated, and autoclaved [19].

◻ **Table 3.2** Some examples of chemical germicides and their functions [7, 17]

Examples of chemical germicides	Function
Chlorine (sodium hypochlorite) (NaOCl)	Fast-acting oxidant, bleaching
Sodium dichloroisocyanurate (NaDCC)	Eliminating blood or other biohazardous liquids
Chloramines	Disinfecting water supplies
Chlorine dioxide (ClO$_2$)	Disinfecting, strong and fast-acting germicide, and oxidizer
Formaldehyde (HCHO)	Eliminating microorganisms and spores
Glutaraldehyde (OHC(CH$_2$)$_3$CHO)	Eliminating Lipid- and nonlipid-containing viruses, fungi, bacteria, and spores
Phenolic compounds	Eliminating vegetative bacteria and lipid-containing viruses, mycobacteria
Quaternary ammonium compounds	Eliminating vegetative bacteria and lipid-containing viruses
Ethanol (ethyl alcohol, C$_2$H$_5$OH) and 2-propanol (isopropyl alcohol, (CH$_3$)$_2$CHOH)	Eliminating fungi, vegetative bacteria, and lipid-containing viruses but not spores
Iodine and iodophors	Preoperative skin antiseptic and surgical scrub
Hydrogen peroxide (H$_2$O$_2$) and peracids	Disinfecting heat-sensitive medical devices

3.6 Laboratory Equipment

3.6.1 General Equipment

General laboratory equipment mainly include pipettes, centrifuges, freezers, incubators, hot plates, coolers, stirrers, water baths, bunsen burners, scales, fume hoods, and also microscopes [18, 20].

3.6.2 Safety Equipment

As aerosols are critical sources of contamination, their dispersion should be reduced. Harmful aerosols can be released by the most of laboratory activities such as mixing, blending, sonicating, grinding, shaking, and centrifuging [21]. Therefore, to achieve contamination control and personnel and product safety, it is seriously recommended that all procedures be performed in BSCs [22]. As a result, not only implementing safety equipment in laboratory is important but also laboratory operator should be trained to ensure safety of equipment regularly. Moreover, it is necessary to implement a system for monitoring equipment calibration [21].

3.6.2.1 Biological Safety Cabinets (BSCs)

BSCs are designed to protect personnel, laboratory environments, and materials against contamination by pathogens. BSCs are classified into three groups based on biosafety levels (◘ Table 3.3) [7, 22].

3.6.2.2 Pipetting Aids, Homogenizers, Sonicators, and Specimen Containers

– Pipettes and pipetting aids [7]:
  – Pipetting by mouth is prohibited. Therefore, personnel must always use a pipetting aid.
  – All pipettes should have cotton caps to reduce contamination.

◘ **Table 3.3** The classification of biological safety cabinets [7]

Types of biological safety cabinets	Biosafety level	Airflow instructions	Protective function
Class I	1, 2, 3	Protective cabinet 0% circulated 100% exhausted HEPA* filter Hard duct	Staff, environment, risk groups 1 and 2 microorganisms
Class II	1, 2, 3	Vertical laminar flow Recirculating air cabinet HEPA filter	As class I plus products, chemicals and risk groups 3 and 4 microorganisms
Type A1		70% circulated 30% exhausted Exhaust to room or thimble connection	
Type A2		70% circulated 30% exhausted Exhaust to room or thimble connection	
Type B1		30% circulated 70% exhausted *Hard duct*	
Type B2		0% circulated 100% exhausted *Hard duct*	
Class III	1, 2, 3, 4	Enclosed glove box with two HEPA filters 0% circulated 100% exhausted Hard duct	As class II and III with complete isolation

HEPA high-efficiency particulate air

— Never blow air into a fluid containing infectious agents, and do not mix infectious agents by blowing and suction.
— Liquids should not be removed by force from the pipette [18, 20].
— Homogenizers and sonicators can reduce small pathogens from liquids and should be used in BSCs or be covered with shields during working [7].
— Specimen containers should be made by high resistance glass, metal, or plastic and should not leak when the cap is in place. No material should remain on the outer surface of the container. It is better to label the dishes properly before filling for easy identification and avoiding hazards [7, 18]. Moreover, specimen containers should be placed on secondary containers such as boxes and racks to prevent their spillage and leakage. Secondary metal or plastic containers should be autoclavable [18].

3.6.2.3 Personal Protective Equipment and Clothing

PPE are important to reduce high risks of contact with aerosols, accidental inoculations, and splashes. Some of the PPEs include the following [23, 24]:
— Laboratory coats and gowns: Laboratory coats protect personnel from workplace hazards such as chemical splashes, chemicals spills, or biological materials such as blood and tissue specimens. Therefore, long-sleeved and fully buttoned coats are better choices. On the other hand, personnel should not wear laboratory coats and gowns when they are outside the laboratory [23, 24].
— Gloves: Latex, vinyl, or nitrile gloves are used to protect against infectious agents, blood, and body fluids [23].
— Face protection devices: Eyes and face protection devices including safety spectacles, safety goggles, face shield, face respirators, and masks protect personnel from hazards that may cause serious injury on the face and eye. Safety goggles and face shield are the best choices to protect staff from chemical splashes, while safety spectacles are not suitable. Moreover, masks with filters are used for protecting against gases, toxic vapors, aerosols, and microorganisms [24].

3.7 Safety Against Fire, Electrocution, and Chemicals

Personnel in microbiological laboratories are exposed to hazardous chemicals as well as pathogenic microorganisms [25]. Therefore, they should be aware of the toxic effects of chemicals, exposure pathways, and potential hazards. Personnel may be exposed to dangerous chemicals through skin contact, needle sticks, ingestion, and inhalation [25]. On the other hand, personnel may face a variety of hazards such as fire, electricity, radiation, and noise. A summary of these hazards and relevant preventive actions is presented in ▢ Table 3.4 [26].

3.8 Biosafety Instruction

Risk assessment process is the basis of biosafety [7]. Risk assessment steps should be carried out by individuals who are well trained and those with the good knowledge of organisms, tools, methods, animal models, and equipment [13]. Accordingly, laboratory director or researcher is responsible for providing risk assessment equipment and facilities

3

◻ Table 3.4 A summary of laboratory hazards, undesirable effects, and ways to prevent them [25, 26]

Types of laboratory hazards	Undesirable effects	Protective measures	Examples
Toxic chemicals	Seriously damaged in the respiratory system, blood, lungs, kidneys, liver, gastrointestinal system, other organs, and tissue. Carcinogenic or teratogenic and skin damage	Attention to chemical incompatibilities	Alkali metals such as cesium, sodium, potassium, and lithium with carbon dioxide, chlorinated hydrocarbons, and water
Explosive chemicals	Fire, skin damage	Attention to chemical incompatibilities	Azides with copper or lead, ethers that have aged and dried to crystals, perchloric acid, picric acid, and picrates
Chemical spills	Fire, seriously damaged in the respiratory system, lungs	Protective clothing such as heavy-duty rubber gloves, overshoes or rubber boots, respirators, chemical spill kits, scoops and dustpans, forceps (for picking up broken glass), mops, cloths and paper towels, buckets, soda ash sodium bicarbonate ($NaHCO_3$) for neutralizing corrosive chemicals and acids, sand to cover alkali spills, nonflammable detergent	Spilled materials and flammable spilled materials
Compressed and liquefied gases	Fire, skin damage	*Compressed gas cylinders and liquefied gas containers*: securely fixed, transported with their caps in place and supported on trolleys, stored in bulk in an appropriate facility at some distance from the laboratory. The area locked and appropriately identified. Should not be placed near radiators, open flames other heat sources, sparking electrical equipment, or in direct sunlight *Small and single-use gas cylinders*: Must not be incinerated	Compressed gas cylinders, liquefied gas containers, and small and single-use gas cylinders

▣ Table 3.4 (continued)

Types of laboratory hazards	Undesirable effects	Protective measures	Examples
Fire hazards	Fire, skin damage	Close cooperation between local fire prevention officers and safety officers, training of laboratory staff in fire prevention, immediate action in case of fire, and using fire-fighting equipment	Electrical circuit overloading, poor and perished insulation on cables, open flames, deteriorated gas tubing, flammable or explosive materials, incompatible chemicals, and sparking equipment
Electrical hazards	Fire, electric shock	Installation of circuit breakers and earth-fault-interrupters in appropriate laboratory electrical circuits	eEectrical installations including earthing or grounding systems
Noise	Hearing problems	Engineering controls such as enclosures or barriers between noisy areas and other work areas or around noisy equipment considering hearing conservation and medical monitoring programs	Certain laser systems
Radiation	*Somatic effects*: leukemia, bone, lung, and skin cancers, minor skin damage, hair loss, blood deficiencies, gastrointestinal damage, and cataract formation *Hereditary effects*: chromosome damage or gene mutation, impaired fertility, menstrual changes in women, congenital malformations, mental impairment, and radiation-induced cancers	Minimizing the time of radiation exposure, maximizing the distance from the radiation source, shielding the radiation source, substituting the use of radionuclides with nonradiometric techniques	Ionizing radiation

3

in collaboration with safety and staff who are responsible for laboratory biosafety. Risk assessment measures should be reviewed regularly and revised as necessary. Revisions will be provided using scientific literatures and other relevant information sources [16].

3.8.1 Assessment of Microbiological Hazards

To identify microbiological hazards, microbiological risk assessment must be performed according to hazard identification, exposure assessment, hazard characterization, and risk characterization [27].

- *Hazard identification*: Hazard identification identifies chemical, biological, and physical agents which can cause an adverse effect on health which is related to the presence of a pathogen in food [27].
- *Exposure assessment*: Exposure assessment provides a qualitative or quantitative estimation of the intake of a microbiological hazard in a specific food or a range of foods [27].
- *Hazard characterization*: The main point in hazard characterization is the association between the amount of exposure to a chemical, biological, or physical agent and the adverse impact on health [13, 27].
- *Risk characterization*: Risk characterization is a combination of hazard identification, exposure assessment, and hazard characterization [27].

3.8.2 The Importance of Proper Documentation

Documentation plays a critical role in the quality system of laboratory which relies on the recording of vital information. Documents should be stored in laboratory and must be available to all laboratory staff. Policies and programs should be followed by biosafety officers. Therefore, every aspect of a new device or a new drug should be recorded, examined, revised, updated, submitted, and archived. In addition, the competency of all staff should be assessed, monitored, and recorded based on qualification and training programs and in relation with the responsibilities [28].

3.8.3 Standard Operating Procedures (SOPs)

SOP is a document which develops based on GLP program. Therefore, SOP content should comply with GLP rules [16]. According to the information obtained from risk assessment processes, biosafety levels, and testing facilities, SOPs are developed to provide the highest level of quality and safety during work [29].

3.8.4 Guidelines for Basic Laboratories: Biosafety Levels 1 and 2

WHO has recognized that biosafety in laboratory is one of the most important international issues. In 1983, WHO published the first edition of the laboratory biosafety manual. Then, codes of practice for safe exposure to microorganisms were developed, and countries were encouraged to enforce these regulations. Since 1983, many countries have applied specific recommendations to develop their codes of practice. Guidelines for basic laboratories (biosafety levels 1 and 2) can be generalized to implement in laboratories with any biological level (▶ Boxes 3.1 and 3.2) [7]. Also, guidelines for containment laboratories (biosafety level 3) and the maximum containment laboratories (biosafety level 4) are developed on the basis of guidelines for basic laboratories [30].

> **Box 3.1 Code of Practice for Basic Laboratories (Biosafety Levels 1 and 2)**
> ▬ Only staff should enter the laboratory.
> ▬ It is necessary to install an international biohazard warning sign on the doors of the rooms with the risk of group 2 microorganisms or higher ones.
> ▬ Laboratory doors should always be closed.
> ▬ There should be special permission for staff to enter animal laboratory. On the other hand, any animal-specific projects are allowed to be entered.

> **Box 3.2 Guidelines for Design and Establishment of Basic Laboratories (Biosafety Levels 1 and 2)**
> 1. Laboratory should have adequate light and avoid any reflection of stunning light and brightness.
> 2. Accumulation and congestion of laboratory equipment, outbreak of insects and rodents, formation of dust (particles), and working with high volume or high concentration of microorganisms should be reduced.
> 3. Laboratory should provide sufficient work area in a safe condition. Cleaning and maintenance should be done.
> 4. Floors, ceilings, and walls should be completely smooth, easy to clean, impervious to liquids and resistant to chemicals and disinfectants. Floors should not be slippery.
> 5. Some facilities outside the workplace should be provided to keep personal clothing, as well as eating and drinking.
> 6. Bench surfaces must be impervious to heat, resistant to water, disinfectants, acids, bases, and solvents.
> 7. In biosafety level 2, autoclave gas sterilizer and other related equipment should be accessible.

3.8.5 Guidelines for Containment Laboratories: Biosafety Level 3

Containment laboratories – biosafety level 3 – are designed to use for procedures involved in group 3 microorganisms or group 2 with a high concentration [7]. This type of laboratories requires more restricted guidelines and rules than basic ones (▶ Box 3.3 [9] and Box 3.4 [31]).

> **Box 3.3 Code of Practice for Containment Laboratories: Biosafety Level 3**
> ▬ Laboratory protective clothing should cover the entire body, and the head and shoe covering should be considered.
> ▬ The biological risk sign should be installed on the main entrance door including information about safety requirements for entering into the laboratory, the biosafety level, and also the name of the laboratory supervisor.
> ▬ Working with infectious agents should be done in BSCs or other containment devices.
> ▬ It is necessary to use respiratory protective equipment.

Box 3.4   Guidelines for Design and Establishment of Containment Laboratories: Biosafety Level 3
Some additional and necessary equipment should be added to basic laboratories to provide a containment laboratory condition:
- A ventilation system should be installed in building. Therefore, air flow from the containment laboratory cannot be recirculated to other spaces within the facilities.
- High-efficiency particulate air (HEPA): Air should be recirculated and reconditioned with HEPA filters within the containment laboratory. Depending on the type of agents, air should be discharged via HEPA filters. When exhaust air is discharged to the outside of the building, it should be dispersed away from air intakes and occupied building.
- Heating ventilation and air conditioning (HVAC): HVAC systems should be installed in containment laboratory to prevent sustained positive pressure. Also, specific alarms should be installed to alert personnel to HVAC system failure.

3.8.6   Guidelines for Maximum Containment Laboratories: Biosafety Level 4

Maximum containment laboratories are designed as a workplace which is involved with group 4 microorganisms. Controlling of these laboratories should be supervised by national health authorities [32] (▶ Boxes 3.5 and 3.6 [7, 32]).

Box 3.5   Code of Practice for Maximum Containment Laboratories: Biosafety Level 4
- Two personnel should work in maximum containment laboratory (two-person rule). Accordingly, no one should work alone.
- It is required to change clothing and shoes completely before entering and upon exiting the laboratory.
- Personnel should be trained in case of emergency such as injury and illnesses and relevant disasters.
- In maximum containment laboratories, a method of communication between personnel should be established and also support personnel outside the laboratory for normal and emergency conditions.

Box 3.6   Guidelines for Design and Establishment of Maximum Containment Laboratories: Biosafety Level 4
Some additional and necessary equipment should be added to containment laboratories to provide maximum containment laboratory condition including primary containment, controlled access, controlled air system, disinfection of effluents, sterilization of waste and materials, airlock entry ports, containment drain, and emergency power.

3.8.7   Guidelines on Laboratory Animal Facilities

Personnel who use animals for laboratory practices and diagnostic purposes should be morally committed to take care of them and avoid any unnecessary harm. Accordingly, adequate food and water, as well as a hygienic and comfortable place, should be provided for animals. For security reasons, animal house should be a unit completely independent of the laboratory. If connected to the laboratory, it should be completely separated from the public parts of the laboratory with perfect disinfectant procedures [33]. Animal labo-

Table 3.5 Containment levels for laboratory animal facilities, methods, and safety equipment [7]

Risk group	1	2	3	4
Containment levels	ABSL*1	ABSL2	ABSL3	ABSL4
Laboratory safety equipment	Limited access, gloves, and protective clothing	ABSL1 methods+ hazard warning signs, biological cabinet Class I or II, disinfection of waste and shelves before washing	ABSL2 methods+ controlled access, biological cabinet, and special protective clothing	ABSL3 methods+ extremely limited access, changing clothes before arrival, biological cabinet class III or positive pressure suits, shower on exit
Cage processing methods	Do not need autoclave	Autoclave cages before washing, changing cages in BSC*	Autoclave wastes prior removing from the area	Autoclave wastes prior removing from the area

*ABSL** animal facility biosafety level, *BSC** biological safety cabinets

ratory equipment can be designated based on a risk assessment protocol, risk group microorganisms, and biosafety levels 1, 2, 3, and 4. A summary of methods and safety equipment used according to animal facility biosafety level (ABSL) can be seen in Table 3.5 [7].

3.8.8 Guidelines for Laboratory Commissioning

The aim of the laboratory commissioning is defining a regular process of monitoring and also collection and verification of documents to ensure that all of the structural components of a GLP-based laboratory are installed, inspected, tested, and approved correctly in accordance with the international or national standards. On the other hand, laboratories are dynamic and complex environments that can be adapted to healthcare needs [28]. All biomedical laboratories should be certified to ensure that they are on the correct way focusing on the following considerations [7]:

- Engineering controls are carried out properly and consistently as they are designed.
- Administrative controls are in accordance with the established protocols and are performed in a suitable site.
- PPE is provided based on established criteria to be suitable for the tasks.
- Materials and wastes are completely decontaminated. Proper waste management practices are carried out quickly.
- For the general safety of the laboratory, including chemical, electrical, and physical safety, there are well-defined procedures which are conducted properly.

3

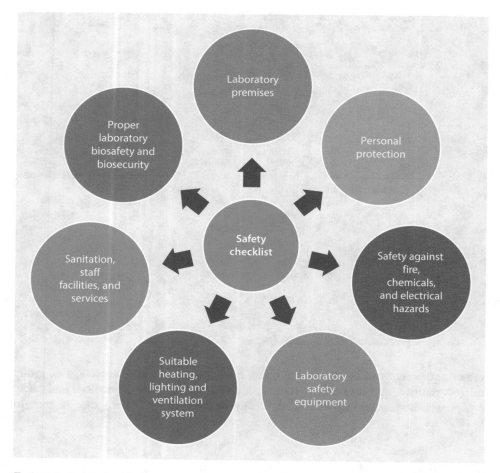

◘ Fig. 3.2 Safety checklist. Important points which should be considered in the safety checklist of GLP [7]. To ensure proper establishment, there should be answer to questions about laboratory premises; staff facilities and services; safety against fire, chemicals, and electrical hazards and also suitable heating, lighting, and ventilation system; and more importantly laboratory biosafety and biosecurity

3.9 Safety Checklist

A proper safety checklist helps director to assess the safety and security of the laboratory. Some important points which should be considered in safety checklist of GLP are indicated in ◘ Fig. 3.2.

3.10 Challenges and Future Perspectives

In recent years, following the advancement in technology and industrialization, biomedical sciences has being dramatically grown. Therefore, to avoid several challenges, appropriate codes of practice and protocols are needed. GLP standards are developed to direct laboratories in a right manner. Accordingly, GLP has been popularized throughout the world. Few people can estimate contemporary challenges about GLP regulatory which is faced today [34]. The aim of GLP is not only decreasing the adverse events of biological products but also improving human health and safety of environment [3]. GLP is also a protocol for nonclinical laboratory researches which can help scientists to perform biomedical researches perfectly. There are several challenges such as issues surrounding record keeping. For example, their reliability and trustworthy especially due to limited access to electronic systems and internet are serious limitations. To overcome these challenges, environmental protection agency (EPA) implements standard requirements that are equipped to receive information electronically for data submission [34]. Other challenges include very old information which is not in accordance with GLP guidelines, heavy costs and regulatory burdens, insufficient staff, and inadequate technical training. Accordingly, there are some procedures that can enhance the quality of the research such as monitoring GLP compliance through regulatory inspections, optimization, and electronic records including signatures, electronic system, internet, etc. [35]. The most important procedure is optimization [36]. It is predictable that other novel methods based on using burgeoning technologies including nanotechnology, biocides, computer modeling, electronic data, and record keeping using the fastest supercomputer with the ability to perform 1000 trillion calculations per second will be developed in the near future [34, 36].

Take-Home Messages

- In 1983, the WHO has developed codes of practice for safe exposure to microorganisms and encouraged countries to enforce these regulations.
- Biosafety is a set of rules that are used to handle the hazards of living organisms and also isolate them in an enclosed laboratory. The base of biosafety is the risk assessment process.
- Risk assessment should be carried out by individuals who are well trained and those with the good knowledge of organisms, tools, methods, animal models, and equipment.
- GPL is a set of techniques that provide safety and quality in the laboratory and environment and also safety of personnel.
- GLP can be used as a practical standard to direct nonclinical laboratory researches.
- Microorganisms are classified into four risk groups based on their possibility of harm in humans, animals, and environments.
- Laboratory facilities are designed and classified based on biosafety levels 1, 2, 3, and 4.
- There are some procedures that can enhance the biosafety and quality control of the laboratories such as using electronic records, internet, optimization, etc.

References

1. Baldeshwiler AM. History of FDA good laboratory practices. Qual Assur J. 2003;7:157–61.
2. Sasaki M, Hinotsu S, Kawakami K. Good laboratory practice (GLP) status of Asian countries and its implementation in non-clinical safety studies in pharmaceutical drug development. J Toxicol Sci. 2009;34(5):493–500.
3. Jena G, Chavan S. Implementation of good laboratory practices (GLP) in basic scientific research: translating the concept beyond regulatory compliance. Regul Toxicol Pharmacol. 2017;89:20–5.

4. Bolon B, Baze W, Shilling CJ, Keatley KL, Patrick DJ, Schafer KA. Good laboratory practice in the academic setting: fundamental principles for nonclinical safety assessment and GLP-compliant pathology support when developing innovative biomedical products. ILAR J. 2018;59:18–28.

5. Wang Q, Zhou WM, Zhang Y, Wang HY, Du HJ, Nie K, et al. Good laboratory practices guarantee biosafety in the Sierra Leone-China friendship biosafety laboratory. Infect Dis Poverty. 2016;5(1):62.

6. Armstrong E. Principles of health and safety and good laboratory practice. In: Remesh V, editor. Biomolecular and bioanalytical techniques: theory, methodology and applications. Hoboken: Wiley; 2019. p. 1–15.

7. World Health Organization. Laboratory biosafety manual. Geneva: World Health Organization; 2004.

8. De Vos FJ, De Decker M, Dierckx RA. The good laboratory practice and good clinical practice requirements for the production of radiopharmaceuticals in clinical research. Nucl Med Commun. 2005;26(7):575–9.

9. Zaki AN. Biosafety and biosecurity measures: management of biosafety level 3 facilities. Int J Antimicrob Agents. 2010;36:S70–S4.

10. Ezzelle J, Rodriguez-Chavez IR, Darden JM, Stirewalt M, Kunwar N, Hitchcock R, et al. Guidelines on good clinical laboratory practice: bridging operations between research and clinical research laboratories. J Pharm Biomed Anal. 2008;46(1):18–29.

11. Brunetti MM. Critical aspects in the application of the principles of good laboratory practice (GLP). Ann Ist Super Sanita. 2002;38(1):41–5.

12. Salerno RM, Gaudioso J, Brodsky BH. Laboratory biosecurity handbook. Boca Raton: CRC Press; 2007.

13. Becker RA, Janus ER, White RD, Kruszewski FH, Brackett RE. Good laboratory practices and safety assessments. Environ Health Perspect. 2009;117(11):A482–A3.

14. Haeckel R. The meaning of good laboratory practice (GLP) for the medical laboratory. Clin Chem Lab Med. 1999;37(2):169.

15. Gouveia BG, Rijo P, Goncalo TS, Reis CP. Good manufacturing practices for medicinal products for human use. J Pharm Bioallied Sci. 2015;7(2):87–96.

16. Worl Health Organization. Handbook: good laboratory practice (GLP): quality practices for regulated non-clinical research and development. Geneva: World Health Organization; 2010.

17. Stevens W. Good clinical laboratory practice (GCLP): the need for a hybrid of good laboratory practice and good clinical practice guidelines/standards for medical testing laboratories conducting clinical trials in developing countries. Qual Assur. 2003;10(2):83–9.

18. Washington JA. Laboratory procedures in clinical microbiology: Springer Science & Business Media. 2nd edition, New York. 1985.

19. Padmanabhan K, Barik D. Health hazards of medical waste and its disposal. In: Energy from toxic organic waste for heat and power generation. Elsevier. Kidlington, United Kingdom 2019. p. 99–118.

20. McGinnis MR. Laboratory handbook of medical mycology. Saint Louis: Elsevier; 2012.

21. Todd CA, Sanchez AM, Garcia A, Denny TN, Sarzotti-Kelsoe M. Implementation of good clinical laboratory practice (GCLP) guidelines within the external quality assurance program oversight laboratory (EQAPOL). J Immunol Methods. 2014;409:91–8.

22. Whistler T, Kaewpan A, Blacksell SD. A biological safety cabinet certification program: experiences in Southeast Asia. Appl Biosaf. 2016;21(3):121–7.

23. Tomas ME, Kundrapu S, Thota P, Sunkesula VC, Cadnum JL, Mana TSC, et al. Contamination of health care personnel during removal of personal protective equipment. JAMA Intern Med. 2015;175(12): 1904–10.

24. Verbeek JH, Ijaz S, Mischke C, Ruotsalainen JH, Mäkelä E, Neuvonen K, et al. Personal protective equipment for preventing highly infectious diseases due to exposure to contaminated body fluids in healthcare staff. Cochrane Database Syst Rev. 2016;4:CD011621.

25. National Research Council. Prudent practices in the laboratory: handling and management of chemical hazards, updated version. Washington, DC: National Academies Press; 2011.

26. Asiry S, Ang L-C. Laboratory safety: chemical and physical hazards. Methods Mol Biol. 2019;1897: 243–52.

27. Hoornstra E, Notermans S. Quantitative microbiological risk assessment. Int J Food Microbiol. 2001;66(1–2):21–9.

28. Gumba H, Waichungo J, Lowe B, Mwanzu A, et al. Implementing a quality management system using good clinical laboratory practice guidelines at KEMRI-CMR to support medical research. Wellcome Open Res. 2019;3:137.

29. Tolba RH, Riederer BM, Weiskirchen R. Standard operating procedures in experimental liver research: time to achieve uniformity. Lab Anim. 2015;49(1 Suppl):1–3.
30. Burnett LC, Lunn G, Coico R. Biosafety: guidelines for working with pathogenic and infectious micro-organisms. Curr Protoc Microbiol. 2009;13(1):1A.1.1–1A.1.14.
31. Day D, Xiang J, Mo J, Clyde M, Weschler C, Li F, et al. Combined use of an electrostatic precipitator and a high-efficiency particulate air filter in building ventilation systems: effects on cardiorespiratory health indicators in healthy adults. Indoor Air. 2018;28(3):360–72.
32. Xia H, Huang Y, Ma H, Liu B, Xie W, Song D, et al. Biosafety level 4 laboratory user training program, China. Emerg Infect Dis. 2019;25(5):e1–4.
33. Feary J, Fitzgerald B, Schofield S, Potts J, Canizales J, Jones M, et al. Evidence based code of best practice for animal research facilities: results of the SPIRAL study. In: 28th international congress of the European-Respiratory-Society (ERS); 2018.
34. Liem FE, Lehr MJ. Future issues including broadening the scope of the GLP principles. Ann Ist Super Sanita. 2008;44(4):335–40.
35. Adamo JE, Bauer G, Berro MM, Burnett BK, Hartman MKA, Masiello LM, et al. A roadmap for academic health centers to establish good laboratory practice-compliant infrastructure. Acad Med. 2012;87(3):279.
36. Kendall G, Bai R, Błazewicz J, De Causmaecker P, Gendreau M, John R, et al. Good laboratory practice for optimization research. J Oper Res Soc. 2016;67(4):676–89.

Further Reading

Books

Armstrong E. Principles of health and safety and good laboratory practice. In: Remesh V, editor. Biomolecular and bioanalytical techniques: theory, methodology and applications. Wiley: Hoboken; 2019. p. 1–15.
World Health Organization. Laboratory biosafety manual. Geneva: World Health Organization; 2004.
World Health Organization. Handbook: good laboratory practice (GLP): quality practices for regulated non-clinical research and development. Geneva: World Health Organization; 2010.

Online Resources

Bolon B, Baze W, Shilling CJ, Keatley KL, Patrick DJ, Schafer KA. Good laboratory practice in the academic setting: fundamental principles for nonclinical safety assessment and GLP-compliant pathology support when developing innovative biomedical products. ILAR J. 2018;59(1):18–28. https://www.ncbi.nlm.nih.gov/pubmed/30566589.
Jena GB, Chavan S. Implementation of good laboratory practices (GLP) in basic scientific research: translating the concept beyond regulatory compliance. Regul Toxicol Pharmacol. 2017;89:20–5. https://www.ncbi.nlm.nih.gov/pubmed/28713068.
Liem FE, Lehr MJ. Future issues including broadening the scope of the GLP principles. Ann Ist Super Sanita. 2008;44(4):335–40. https://www.ncbi.nlm.nih.gov/pubmed/19351991.

Design of Experimental Studies in Biomedical Sciences

Bagher Larijani ⓘ, *Akram Tayanloo-Beik* ⓘ, *Moloud Payab* ⓘ,
*Mahdi Gholami, Motahareh Sheikh-Hosseini,
and Mehran Nematizadeh*

© Springer Nature Switzerland AG 2020
B. Arjmand et al. (eds.), *Biomedical Product Development: Bench to Bedside*,
Learning Materials in Biosciences, https://doi.org/10.1007/978-3-030-35626-2_4

4.1 What You Will Learn in This Chapter

Proposing, investigating, and testing new theories lead to a considerable progress in science. In this concept, appropriate experimental design has a fundamental importance. Well-designed experimental studies with considering key points which are appropriately analyzed and reported could maximize scientific gains. However, there are still some problems in this process that should be addressed. Accordingly, using adequate methods based on the aim of experiments, qualifying the laboratory tests, pointing reproducibility of results, considering statistics for data analysis, reporting data transparently, and conducting pilot studies are fundamental considerations that will be discussed in this chapter. This chapter begins with a brief definition of experimental studies and different types of research in biosciences. Consequently, general principles of the experimental study design and conduct will be reviewed. Thereafter, some limitations and challenges in the field of experimental studies and also validation and standardization will be described.

4.2 Rationale and Importance

Conducting an experimental study requires considerable time and resources. In order to balance between costs and benefits and to achieve reliable, valid, and reproducible results, the study must be designed appropriately and focusing on a standard and scientific method according to experimental purposes. Laboratory tests should be validated and qualified, and suitable statistical analysis is needed in a scientific setting. In other words, using inappropriate and limited resources may reduce validity, and the power of the study with more and more consumption leads to resource wasting in addition to ethical problems. Moreover, a well-designed study is later can be published in a higher impact journal which is available for a huge number of professionals. As a result, sufficient data should be reported perfectly. Accordingly, published data can be used for designing and conducting future investigations. As a conclusion, a proper study design can almost ensure its validity and usefulness [1–3].

4.3 Experimental Studies in Biosciences

Medical research methods are divided into two major types, including primary and secondary researches. A secondary research focuses on using the data from previous studies and includes reviews and meta-analyses. A primary research, on the other hand, is the actual data gathering through experimental, clinical, and epidemiological surveys. Animal, cell, biochemical, physiological, and genetic studies are examples of experimental studies [4]. As it can be inferred from the word "experiment," these studies are usually conducted in controlled situations, considering important variables. These controls help the investigator to measure and estimate quantities and test the probability of different hypotheses, using methods that seem impossible or very different from the ones used in real life. For example, in an experimental study, estimations and procedures focused on a sample are named study populations, while most of the target population is not directly included in the measurements. In accordance with this, the sampling procedure plays an important role in experimental studies. Results obtained from them are generalized to the target population to make more appropriate decisions and conclusions. However, inevitable varieties of samples and population cause uncertainties in study results. In other words, if the whole population had been studied, the results would have been different totally. Uncertainty is more prominent when the study is conducted on different species. In animal studies, study population includes humanized animal models to resemble controlled medical conditions relevant to human being. After designing appropriate animal models, research hypothesis can be tested. For instance, two different therapeutic processes can be compared. Each group of animal models is treated with one method, and after that, disease markers or other proxies showing disease activity are measured and compared between groups. There is other type of experimental study which is designed to understand the dose response of various medications [5]. Furthermore, some studies emphasize on examining enzymes, markers, or genes, evaluating imaging techniques such as computed tomography (CT) and magnetic resonance imaging (MRI), gene sequencing, and finding correlations between genotypes and phenotypes. Statistical test development and modeling are also considered as experimental studies. Overall, different studies are designed for different purposes in basic research [4–6].

4.4 Principles of Experimental Study Design

As it was discussed previously, each experimental study has its specific goals and objectives and must be designed accordingly. A well-designed study should be able to detect scientifically, while being simple and bias free. To achieve these goals, key steps in study design must be followed (◘ Fig. 4.1).

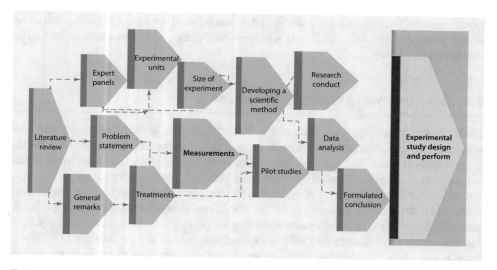

◘ **Fig. 4.1**   Flowchart of experimental study design. There are some principles that need to be considered in design and performance of experimental research. Each of them has a key impact on the study design procedure [7–10]

4.4.1 **General Remarks**

Regarding the subject of the study, it is obvious that the final goal in every research is to solve a problem, answer to an important question, or further progress toward goals of previous studies. As a result, a beneficial study targets important phenomena and is interesting enough for persuading responsible organizations to provide needed resources. A study may be a part of bigger investigation plans, industrial endeavors, or efforts made to treat an unresponsive disease as long as it has its own benefits.

4.4.2 **Expert Panels**

Collaborators and stakeholders have a determinant role in validity of different studies. Therefore, before starting practical procedures, probable needs for experts in different fields should be evaluated. As a necessity, members of expert panel are gathered to develop the study policies and required methodologies [1].

4.4.3 **Literature Review**

First of all, scientific data must be gathered from different sources such as journals, textbooks, and internet, to form a base of evidences for facilitating study development. A perfect literature review is necessary for learning about previous studies relevant to the intended subject. This information is helpful in finding suitable animal models and prevents testing-approved hypothesis. Actually, it helps scientists to avoid reinventing the wheel.

4.4.4 **Problem Statement, Objectives, and Hypotheses**

It is important to clarify what the problem is, the possible solution, and how it would be tested. Significance of the targeted problem should justify the needed costs and efforts. Experiment objectives are main goals of the whole process that should be developed using related questions and tests. Hypotheses, possible final results of the tests, are defined in two main categories. Null hypothesis is correct when there is no difference between examined groups. In contrast, alternate hypothesis is accepted if a statistically significant difference is present. Problem statement and hypotheses may change during the investigation, but are essential for starting a beneficial study. For example, if the study objective is to determine whether drug A or B is more effective in increasing ejection fraction in mice, alternate hypothesis states that the mice treated with either drug A or B will have significantly higher ejection fractions compared to the other group. Null hypothesis, on the other hand, says that no significant difference will be observed among experiment groups. Finally, in creating hypotheses, terms such as "better heart function" lead to untestable ones that are not defined clearly and should be avoided.

4.4.5 **Experimental Units**

Experimental units are physical entities that can be given randomly determined treatments in an experiment, and all changes and outcomes are analyzed statistically. As experimental units can vary significantly among different research fields, general points will be focused on in this section. Units should be chosen in an evidence-based setting to provide more clarified conclusions and applicable results to the target population. Moreover, depending on the objectives, scientists can decide on the diversity of experimental units. Divergent study groups are usually harder to be investigated but may be more advantageous in some circumstances. For example, if the results are going to be used for the target population, this variety should be considered in the sampling process. However, subjects such as examining an unknown process need homogenous samples. Even in some cases, the same unit is used several times and receives different treatments during a specific period and is somehow a control for itself. This is an example of unit changes over time and brings forward the importance of differentiation between units and observations. The key point that helps here is the fact that different units must be able to receive different treatments. For instance, the cage of a group of animals that receive a treatment through their identical diets in the same cage is considered the experimental unit. In such situations, if differences among individuals in the same unit are important for the researcher, it must be assumed and stated that individuals would have been in their current state even in different cages or units. Overall, in a series of interrelated studies, a combination of these approaches may come in handy, and this obviously affects the statistical analysis process [6].

4.4.6 **Treatments**

Even in a simple study, comparison between two or more treatments is the main goal. Treatments, T, are compared with each other and with the control group, C. Treatment and control groups must be précised in every possible feature, to make inferences more reliable and valid. Different degrees of similarity may be present between control and treatment groups, depending on the variables that are going to be compared. Examples of these controls can be placebo, a known treatment, a pharmacologically inactive agent, or even no treatment at all. Generally, ethical rules have a crucial role in choosing control groups and also designing study methods.

4.4.7 **Measurements**

Which parameters are going to be measured? Are they reliable markers for disease activity or drug effectiveness? These are two fundamental questions which to be answer. Other than that, errors in measurements aim to control them are critical to the research process. Generally, in every experiment, some relevant variables must be tested statistically. Variables are divided into three main categories. (1) Baseline variables show baseline statement of participants. (2) Response variables measure the final response. (3) Intermediate variables bridge between baseline and response ones. Additionally, intermediate variables can clarify the process that leads to the response and help scientists detect early stages of final responses or even serve as proxies for the main response. As it seems obvious, variables that are going to be used must be specified in accordance with the study purposes.

4.4.8 **Sample Size**

In addition to several principles of study design, determining an appropriate sample size is crucially important. As the sample size increases, the study becomes more difficult to conduct, but the results are more valid and reliable. Accordingly, costs and validity of the study should be balanced. For instance, in some cases, using tests with low precision leads to a large sample size and serious difficulties in performing the study.

4.4.9 **Developing a Scientific Method**

Method development is critical for designing an experimental study and includes four major aspects. (1) The scientific problem statement must be defined. (2) Problem solution must be found regarding the hypothesis. (3) Possible consequences of hypothesis must be predicted. (4) Hypothesis must be tested by conducting the research procedures. These form the method of the investigation.

4.4.10 **Pilot Studies**

In order to design an experiment for large samples, some basic data such as means or standard deviations are necessary. This information may be available in a body of literatures. If not, pilot studies will help scientists provide them.

Small experiments are performed, and the results concluded from experiment study are used for designing main study procedures and calculating sample size. Because of these considerations, pilot studies should be large enough to achieve reliable information [6].

4.4.11 **Conduct the Research, Analyze Data, and Formulate Conclusions**

In this stage, the designed experimental procedure will be conducted and properly analyzed. Finally, a formulated hypothesis should result to confirm the scientific concept.

In conclusion, a well-designed study is planned, described, and conducted, in a way that makes it possible to be repeated in other situations. On the other hand, objectives and hypotheses must be defined precisely, and sample size must be explained carefully. Unfortunately, in some cases, these steps are not followed correctly and result in serious problems that make the study process difficult.

4.5 **Common Concerns in the Design of an Experimental Study**

In spite of all efforts, there are some obstacles in the field of experimental studies. These include problems in reproducing the studies and results, defective data analysis, errors in randomization and blinding, and the reporting process. Additionally, feasibility of the study and costs are the rest of obstacles.

4.5.1 **Feasibility**

Before planning for an experimental study, researchers should know whether they are able to conduct the research or not and make sure that their intended methods produce reliable results. Accordingly, pilot studies may not produce exactly the desired result of the investigation, but provide lots of help in such situations and give researchers valuable information about the needed time, subjects, and resources. This helps investigators choose the method with the least needs. However, pilot studies are not always taken seriously, and there is still controversy in their usefulness for every study that is going to be designed.

4.5.2 **Reproducibility of Results**

Reportedly, 51–89% of published researches cannot be reproduced in other research sites. It is a key concern among many research-related organizations all over the world. For example, in preclinical studies, it has become a part of the reduction principle of the 3Rs as it may ruin study validity. Ethical problems and time and money wasting are some consequences of this phenomenon. It is not negligible, all the researches should be designed in a reproducible setting with special attention to method development, and the following concerns should be paid attention to.

4.5.3 **Vulnerability in Statistical Design**

A good statistical design can help the researchers discover more, despite conducting less experiment. It can also improve study validity through mechanisms such as replication, randomization, and blinding. Unfortunately, not many researchers use statistics to improve cost-effectiveness and validity of their research protocols especially in animal studies.

4.5.4 **Data Reporting**

If a study is meant to be useful, its result should be reported properly. Details such as sample size calculation techniques, study limitations, randomization process explanation, inclusion and exclusion criteria, etc. must be mentioned correctly to ease evaluating and reproducing. Designs based on incomplete reports may even have harmful effects on participants. Further, researchers might be willing to report positive results only, and this is another problem in data reporting.

4.6 **Validation of Designed Study**

If it has been successful in examining the intended hypothesis, then its results can be reliable; therefore, a study is valid. Experimental validity is also related to study generalizability to a larger population and is explained in two groups: (1) internal validity and (2) external validity.

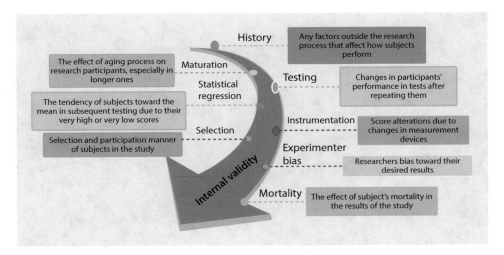

Fig. 4.2 Internal validity threats. Other than statistical regression, instrumentation, selection and experimenter bias, and mortality, history, testing, and maturation are eight major factors that can reduce the study validity and need to be considered [10–12]

4.6.1 **Internal Validity**

If a study is able to prove or deny a causal relationship between one or more independent variables and one or more dependent variables confidently, the study is internally valid. In fact, it should make researchers sure that the results are exact effects of the tested treatments. As a major determinant of research value, validity should be maintained. The following are some threats that should be considered in the internal validity of the study (■ Fig. 4.2).

4.6.2 **External Validity**

External validity shows the extent of generalizability conducted study. It actually means that the study results represent a larger population's characteristics with reasonable certainty. Studies may show reasonable significance level in their results, but threats such as demand characteristics, Hawthorne effects, order (carryover) effects, and treatment interaction effects may cause irrelevance between study results and real features of the target population (■ Fig. 4.3).

It is almost impossible to control all threats to study validity. However, researchers must try their best. Finally, good randomization, considering as much external factors as possible, and planning in advance are giant steps to a valid research.

4.7 **Challenges and Future Perspectives**

Globally, experimental researches are considered as one of the most suitable ways of investigating different hypotheses. In spite of several benefits, there are some limitations and challenges in experimental studies. One of them in such investigations is artificial condi-

4

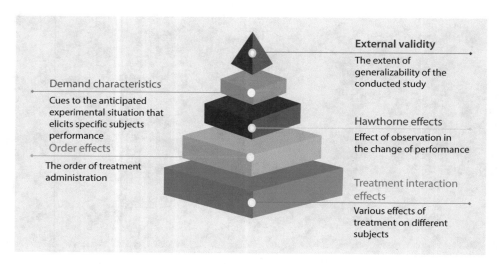

Demand characteristics

Cues to the anticipated
experimental situation that
elicits specific subjects
performance

Order effects

The order of treatment
administration

External validity

The extent of
generalizability of the
conducted study

Hawthorne effects

Effect of observation in
the change of performance

**Treatment interaction
effects**

Various effects of
treatment on different
subjects

◘ **Fig. 4.3** Threats to external validity of research. External validity can be threatened by some factors
that lead to a reduction in the generalizability of the study. Results of ignoring these four classes of
threats can extend to the larger study population [10, 13, 14]

tions that create for study which is totally differed from a real situation, because control
condition of experimental study can be provided by restricting tremendous environmental effects. Therefore, in order to overcome this limitation, validation of designed study
according to guidelines and ethical standards can be helpful. Additionally, uncontrolled
extraneous variables, human errors, incomplete data reporting, the vulnerability in statistical design, and high costs are considered as other limitations in design and performance
of experimental research. Hence, the selection of a correct type of study and specified
design and precise implementation lead to minimize a lot of challenges.

Take-Home Messages

- Experimental studies make the foundation of sciences.
- Experimental study is a multidisciplinary type of research that helps to deter-
  mine different causation.
- An appropriate study design based on the purpose of experiment causes more
  reliable results.
- Pilot studies can provide lots of information about the feasibility of the study.
- A proper statistical design can improve the validity and cost-effectiveness of the
  study.
- The proper data reporting is a significant factor that facilitates the evaluation of
  the study and empowers reproducibility features.
- Generalizability of a study is strongly related to the appropriate experimental
  validation.

References

1. Johnson PD, Besselsen DG. Practical aspects of experimental design in animal research. ILAR J. 2002;43(4):202–6.
2. Festing MFW, Overend P, Cortina Borja M, Berdoy M. The design of animal experiments: reducing the use of animals in research through better experimental design. Revised and updated ed. London, UK: Sage Publications Ltd; 2016.
3. Kilkenny C, Parsons N, Kadyszewski E, Festing MF, Cuthill IC, Fry D, et al. Survey of the quality of experimental design, statistical analysis and reporting of research using animals. PLoS One. 2009;4(11):e7824.
4. Röhrig B, du Prel J-B, Wachtlin D, Blettner M. Types of study in medical research: part 3 of a series on evaluation of scientific publications. Dtsch Arztebl Int. 2009;106(15):262.
5. Parker RM, Browne WJ. The place of experimental design and statistics in the 3Rs. ILAR J. 2014;55(3):477–85.
6. Festing MF, Altman DG. Guidelines for the design and statistical analysis of experiments using laboratory animals. ILAR J. 2002;43(4):244–58.
7. Thorndike E. Animal intelligence: experimental studies. London: Routledge; 2017.
8. Creswell JW. Steps in Conducting a Scholarly Mixed Methods Study. DBER Speaker Series; 2013. 48.
9. Roy RK. Design of experiments using the Taguchi approach: 16 steps to product and process improvement. New York: Wiley; 2001.
10. Yu C-H, Ohlund B. Threats to validity of research design. 2010. Retrieved January 12, 2012.
11. Henderson VC, Kimmelman J, Fergusson D, Grimshaw JM, Hackam DG. Threats to validity in the design and conduct of preclinical efficacy studies: a systematic review of guidelines for in vivo animal experiments. PLoS Med. 2013;10(7):e1001489.
12. Christensen LB, Johnson B, Turner LA. Research methods, design, and analysis. Boston: Pearson; 2011.
13. Ferguson L. External validity, generalizability, and knowledge utilization. J Nurs Scholarsh. 2004;36(1):16–22.
14. Brewer MB, Crano WD. Research design and issues of validity. In: Handbook of research methods in social and personality psychology. Cambridge: Cambridge University Press; 2000. p. 3–16.

Further Reading

Articles

Bernard HR, Bernard HR. Social research methods: qualitative and quantitative approaches. Thousand Oaks: Sage; 2013.

Fraenkel JR, Wallen NE, Hyun HH. How to design and evaluate research in education. New York: McGraw-Hill Humanities/Social Sciences/Languages; 2011.

Gall MD, Borg WR, Gall JP. Educational research: an introduction. New York: Longman Publishing; 1996.

Keppel G. Design and analysis: a researcher's handbook. Englewood Cliffs: Prentice-Hall, Inc; 1991.

Punch KF. Introduction to social research: quantitative and qualitative approaches. London: Sage; 2013.

Books

Fan J. Local polynomial modelling and its applications: monographs on statistics and applied probability 66. Boca Raton: Routledge; 2018.

Festing MF, Altman DG. Guidelines for the design and statistical analysis of experiments using laboratory animals. ILAR J. 2002;43(4):244–58. https://doi.org/10.1093/ilar.43.4.244.

Quinn GP, Keough MJ. Experimental design and data analysis for biologists. Cambridge: Cambridge University Press; 2002.

Preclinical Studies for Development of Biomedical Products

Mehran Nematizadeh, Moloud Payab (iD)*, Mahdi Gholami,*
Babak Arjmand (iD)*, Bagher Larijani* (iD)*, and Akram Tayanloo-Beik* (iD)

© Springer Nature Switzerland AG 2020
B. Arjmand et al. (eds.), *Biomedical Product Development: Bench to Bedside,*
Learning Materials in Biosciences, https://doi.org/10.1007/978-3-030-35626-2_5

5.1 What You Will Learn in This Chapter

Clinical research progressions are highly dependent on answering this question that how different factors such as drugs, chemicals, microorganisms, or hormones affect the human body? Unfortunately, it is not usually possible to address these questions through performing relevant tests on human population. Hence, developing animal models which properly mimic the human pathophysiology to generate different human disease situations is a suitable alternative. However, there are multiple stages for translating basic researches and moving from bench to bedside. For instance, one of the most crucial stages in developing safe and effective pharmaceutical or cell-based products is to design an appropriate preclinical study for demonstrating its safety and efficacy and applicability. In this regard, providing an appropriate and valid animal model seems to be critical in the development of novel treatments. Accordingly, this chapter has a special focus on preclinical study principles, essentials, and its translational importance in bioscience. At the beginning of this chapter, the general definition of preclinical studies and its rationale and importance will be discussed. Then, some of the essentials that should be concerned in preclinical studies will be presented, and in the following, an animal model will be defined, and translational importance of animal modeling will be argued. At the end of this chapter, several validation strategies in the developed model will be addressed according to relevant guidelines and standards.

5.2 Rationale and Importance

The development of more appropriate basic studies places a great impact on clinical research findings. In fact, it is a fundamental stage for bridging translational gap on moving from the basic research to the clinic. Therefore, designing experiments called preclinical studies, which use animal models for not only predicting probable results of interventions but also evaluating their safety and feasibility, is necessary [1]. Despite fewer restrictions, dealing with animals requires sufficient knowledge about the field so that reliable data is produced with the least possible harms and costs [2]. Following animal studies, the results should be processed for use in clinical studies. However, not all preclinical studies lead to applicable clinical ones. For example, many drug development processes stop in clinical application due to unpredicted adverse effects and the lack of reasonable effectiveness. To minimize these surprises, appropriate animal models are of critical importance. Moreover, those new discoveries made in clinical trials can be used for further improvements in animal models. In conclusion, properly designed animal studies lead to reproducible clinical results that bring improvements in bedside and benchside medicine [3, 4].

5.3 Preclinical Study Essentials

Other than animal modeling, an important determinant of study validity, many factors affect outcome of the research. Most of them are related to the practical procedures and methods for controlling possible threats. In spite of their importance, the following points are not considered seriously.

5.3.1 **Duration of Treatment**

It is critical to pay attention to the time gap between disease induction and treatment initiation. Real diseases are often diagnosed after establishment when they have led to specific clinical manifestations. However, animal models are usually treated immediately after induction in the early stages of the pathologic process. In fact, the tested treatment serves as a prophylaxis rather than a therapeutic treatment, and this may lead to an overestimation of the effect.

5.3.2 **Animal Compatibility**

Interactions between animal models and other factors in a research must be taken into account when choosing an animal model. The target drug must be pharmacologically active in the selected animal, and demographic variables and biologics should have a proper relation with the intended medical condition and also human physiology. For instance, some disorders such as Parkinson and Alzheimer's disease (AD), stroke, osteoarthritis, arthritis, and cancers are almost specific to the elderly [5–7]. As a result, it is better to use older subjects to test those treatments. Moreover, some characteristics are only present in primates and therefore cannot be tested in other animal models.

5.3.3 **Blinding and Measurement Bias**

If a possible result is favorable for the researchers, measurements might become affected. Despite all efforts to control personal preferences, these may change the way different parameters are measured and reported, especially when there are subjective variables in the study. Proper blinding of the personnel, responsible for measurements, however, is able to reduce this bias considerably.

5.3.4 **Reproducibility of Results**

As it was discussed previously, an animal study, similar to others, should be conducted and reported in a way that it can be repeated with almost the same results in another site or with slightly different environmental parameters. Strict standardizations or incomplete reporting may largely contribute to this problem. Lots of preclinical studies never lead to considerable clinical applications because of irreproducibility. Accordingly, a large number of animals are harmed, but the results cannot be used in clinical fields. Some scientists blame this condition and state that we might be able to get more reproducible results without increasing the sample size, if we combine the data about almost the same subjects from at least two different sites and try to conclude from the whole package.

5.3.5 **Sample Size**

Minimum sample size favorably considers the minimums for conducting a valid experiment. Regarding the 3R principle, introduced by Russel and Burch, animals should be used only when there is a real necessity (Replacement), the minimum possible numbers of animals should be used (Reduction), and, finally, minimum animal suffering should be considered during the study (Refinement).

5.3.6 Personnel

Personnel involved in the research process and in contact with animals should also meet some standards, according to regulations on animal ethics and welfare. Moreover, they should be trained to achieve related expertise. In this regard, even though the project managers may not provide the training themselves, they should ensure it has been given to the personnel.

5.3.7 Data Reporting

Proper data reporting in animal studies makes it possible to be reproducible. In this regard, it is important to report both positive and negative results.

While preclinical studies bring clinical improvements, clinical findings can also be used to direct future preclinical studies. For example, if a new drug, manufactured with the help of proteomics, translation, and back translation, fails to produce the desired effect, we can go one step back. Therefore, through preclinical studies, problems can be found and resolved [8].

5.4 Animal Models

Models are simple representatives of more complex subjects that are also usually more accessible and easier to deal with. Ethics strictly ban interventions with doubtful consequences on humans and risking human lives. As a result, animal models, in biomedical research, are often used instead of human beings for evaluating new treatments for different diseases or injuries that are present or induced. Animals were first used in ancient Greece for observations that helped them understand human anatomy and physiology. An ideal animal model is the one with the least differences and the most similarities with human characteristics. From single-celled organisms to the most evolved animals, this similarity increases when the animal is phylogenetically closer to human. However, other important factors such as sensitivity to the drug, availability, and the disease which is going to be studied also affect the selection of an appropriate animal model. For example, motor system and behavioral investigations are usually conducted on monkeys, due to their behavioral similarity with humans. On the other hand, mice are suitable for cellular and molecular investigations, and even non-mammalian animals are used when gene expression is going to be investigated. Three main types of animal models, classified according to the extent of similarity in causes, symptoms, and treatment options, are (1) homologous, (2) isomorphic, and (3) predictive models. Homologous animals are the ones with similar causes, symptoms, and treatment options compared with a specific human condition. Isomorphic models have the same symptoms and treatments, and predictive animal models only share treatment characteristics. This kind of model is used when there is no identified cause for the target disease [9, 10].

Another classification divides animal models into five groups: (1) induced (experimental) models, (2) spontaneous models, (3) genetically modified models, (4) negative models, and (5) orphan models. Induced animal models are previously healthy animals that are manipulated in laboratory settings. The process induces a condition similar to a known human condition in animal model. Streptozocin-induced diabetes and strokes caused by surgical interventions are common examples. Spontaneous animal models are natural products of mutations that occurred in normal situations. Many different types of spontaneous rodent models are now available after they have been accidentally identified. These include models for hypertension, arthritis, and diabetes. Genetically, modified disease models include animals with foreign extra DNA and also animals deprived of specific genes by genetic engineering methods. Spontaneous and genetically modified animal models are called transgenic and knockout models, respectively. These models are best options for examining the effect of genes on disease process and also their role in susceptibility or resistance to a specific medical condition. Further, some animal models are resistant to specific diseases. Therefore, evaluating the underlying cause may provide useful clues to understand the disease pathophysiology. Models used in this type of animal study are called negative models. Finally, orphan animal models are ones with conditions that are not properly described in humans [11]. After gathering data about different types of animal models for a specific disease, a delicate choice can be made considering the study goals and available resources (◼ Table 5.1). In the next step, the selected animal model should be validated according to the criteria provided to evaluate animal models.

5.5   Translational Importance of Animal Models

As it was discussed, no significant manipulation is allowed in human treatment procedures, unless it is guaranteed with the help of preclinical studies that the procedure is at least safe for human subjects. In fact, safety is essential even for obtaining informed consent from patients. Despite all these, no preclinical test can be used in clinical studies without necessary and adaptive modifications, which make it suitable for human physiology. However, the goal is to make this gap as small as possible. This is simply the reason for the use of live animal models instead of in vitro investigations that do not take complex interactions inside a viable body into action. Animal studies also help us understand the pathophysiology of human diseases [26, 27]. Another positive feature of animal models is their ability to reproduce new generations with the least variations and sufficient numbers, while having different features among different types of them. Different disease types such as psychiatric, neurodegenerative, and neurodevelopmental ones need various kinds of animal models. Lesioned, transgenic, knockout, and selective breeding models are more commonly used according to the disease that is going to be mimicked [28]. Although animal models are quite helpful in medical research, sometimes their results are not reproducible in clinical trials. These unexpected happenings also raise economic and ethical questions. If clinical trial failures are going to be reduced, animal studies should be properly designed, enough numbers of animal subjects should be used, and interventions should be delivered with full control [26].

Table 5.1 Some examples of the humanized animal models. Several human disease conditions have been modeled in different animal species. According to specimen characteristics and the purpose of the study, the appropriate animal model is determined [12–26]

	Animal Species	Human modeled disease	Advantages	Limitations
1	Monkey	Parkinson's disease (PD), cognitive impairment, Huntington's disease (HD), (AD), malaria	Physioanatomical and behavioral similarity with human Valuable models for the study of malaria pathogenesis Suitable for the study of pregnancy mechanism and hormones due to similarity with human Beneficial for fundamental PD and tremor studies Suitable for the study of auditory cortex	Limitation in widespread use due to ethical issues Difficult accessibility High cost Difficult handling
2	Marmoset	PD, cognitive behavioral impairment, AD, genome editing	Easily handling because of small size A high rate of reproductivity Common social characteristics with humans Useful for studying behavioral and cognitive disorders	Unsuitable for positron emission tomography scans (PET) Fail many cognitive ability tests
3	Mouse	PD, hypertrophy, experimental autoimmune encephalomyelitis (EAE), Type 2 diabetes, Type 1 diabetes, HD, AD, retinal disease, cancer	Similar genome to human (99%) Available genetic and molecular data Easy performance of large-scale studies because of small size Cost-efficient Available transgenic models Easy to handle	Different biomechanical environment with humans Inapplicable some of small scale defects in human condition
4	Rat	PD, heart failure, hypertrophy, EAE, Type 2 diabetes, Type 1 diabetes, cognitive impairment, AD, cartilage defects	Possibility of a decrease in biological variation Easy to maintenance Suitable bridge between in vitro and in vivo studies Proper for proof of concept data Cost-efficient Easy to maintenance Availability Easy to handle	Different biomechanical environment with humans Inapplicable some of small scale defects in human condition

5

◘ **Table 5.1** (continued)

	Animal Species	Human modeled disease	Advantages	Limitations
5	Cat	PD, heart failure, hypertrophy, AD, retinal disease, HIV (human immunodeficiency viruses), AIDS (acquired immune deficiency syndrome), and their counterparts, cancer	Long life span Genetic diversity in population Naturally infected with FIV virus Suitable for vision studies Usually affected by heart disease in lifetime	Limitation in widespread use due to ethical issues
6	Pig	PD, heart failure, hypertrophy, AD, cartilage defects	No ethical concern Similar physiological parameters with humans Similar biochemistry to human Similar bone apposition rate and trabecular thickness to human	Expensive maintenance for long-term studies Specialized habitats
7	Dog	Type 1 diabetes, PD, heart failure, hypertrophy, canine cognitive dysfunction syndrome (CCDS), retinal disease, cancer, van den Ende-Gupta syndrome (VDEGS), Raine syndrome, cartilage defects	More than 360 common diseases with human Spontaneous model for heritable human diseases Unique structure of population High similarity with human genome Available genealogical and veterinary data Suitable model for rare disorders of human	Limitation in widespread use due to ethical issues Complex ethical approval process
8	Sheep	PD, heart failure, hypertrophy, HD, AD, cartilage defects, genetic disease, cystic fibrosis (CF), respiratory diseases	No ethical concern Easily available Low maintenance cost Large brain size Similar anatomy to humans Suitable model for surgical trials Similar respiratory system to humans	Difficult to handle
9	Ferret	Hypertrophy, respiratory diseases, influenza	Same respiratory viruses as humans Similar lungs and airways bare with human and similar clinical pathology and immunity	Expensive maintenance

(continued)

◻ **Table 5.1** (continued)

	Animal Species	Human modeled disease	Advantages	Limitations
10	Horse	Cartilage defects, osteoarthritis, type 2 diabetes, obesity, uveitis, cancer, respiratory diseases, Equine metabolic syndrome (EMS)	Similar large defects with human Large joint surface Suitable for cartilage defect study and long-term follow-ups Suitable for studying prediabetes phase for long time	High cost Specialized center Require
11	Zebra-fish	PD, cancer, drug discovery, infection and inflammation, muscle and cardiac disease, neural disorders, skeletal disease, toxicology, hematological disorders, gastrointestinal metabolism disease	Similar physiological and structural characteristics with human Orthologous genes and proteins with human Suitable for large-scale random mutagenesis screens Extensive manipulation of gene activity Suitable for drug treatment investigation	Not mammals Lack of placenta

5.6 Validation Strategies of Preclinical Studies in Biomedical Sciences

An animal model is considered valid if it shares enough similarities with the human pathophysiologic condition. It is going to represent etiology, pathophysiology, symptomatology, and response to therapeutic interventions. Validity is evaluated according to the criteria as follows:

— *Face validity:* Face validity is achieved when model biology and symptoms are similar to the actual disease. In spite of the importance of this aspect, insufficient data about disease pathophysiology makes it difficult to assess face validity.

— *Predictive validity:* Predictive validity is the model's ability to predict similar clinical effects from the same tested interventions. As disease mechanisms are not exactly the same across different species, and drugs designed for humans cannot produce exactly the same effects in animals, predictive validity is another challenging criterion for studies to be fulfill.

— *Target validity:* Target of the medications tested in the study should have similar roles in the animal and human physiology. For example, beta-3 adrenergic receptor is an important factor in energy metabolism in rodents, while does not have this role in humans.

The mentioned validation criteria check model validity in different fields and for different purposes. As a result, it is important to choose an animal model which is more validity in the fields that are main concerns of the desired study. In other words, animal models should be fit for the purpose. On the other hand, each aspect of model validity becomes more important in a specific condition. For instance, if a drug is going to be used in clinical trials in near future, predictive validity is critical to researchers when studying its effects. While there is no perfect animal model to be used as a gold standard, some guidelines have been developed to help researchers improve the quality of designed animal models and studies [25, 29]. In the following section, some features of these guidelines will be discussed.

5.7 Guidelines and Standards for Preclinical Studies

In order to run an animal study, guidelines must be followed for ensuring rational use of animal models during the study process. Many organizations, at institutional or national level, evaluate the application of these guidelines and animal welfare requirements in experimental animal studies. As animal protection groups and public opinions became concerned about animal well-being while they are being used in studies, these organizations were established and relevant guidelines were developed [30]. Moreover, ethical consideration has become important in experimental studies since the nineteenth century. The first society dedicated to animal protection was called the Society for the Preservation of Cruelty to Animals and was founded in England in 1824. After, in 1860, the wife of a physiologist created the first association for defending laboratory animals. This was because Claude Bernard, her husband and the father of contemporary experimental physiology, used their daughter's pet for showing different tests in his classes. The British Cruelty to Animal Act, the first law for regulation of animal use in investigations, dates back to 1876 in Great Britain. However, it was not earlier than 1909 that such regulations for ethics in experiments were proposed in North America. Also, during the 1980s, a movement was formed to completely ban the use of animals in biomedical research in the United States, Britain, Canada, and Australia and finally led to a declaration on the use of animals in research by the World Medical Association. Already, the European Directive 2010/063/EU of September 23, 2010 laws control animal researches across Europe, and studies involving animals must be approved by regulatory bodies [31]. When an animal study is being planned, the mentioned regulatory guidelines must be paid attention to, as ethical requirements for starting the practical phase of the study. Other than reducing the number of animal models used, following these guidelines will also help assess the data from previous similar studies and provide better test methods accordingly. However, limitations in regulating animal testing in advanced studies are inevitable and should be discussed with regulatory bodies. Among the guidelines available, the most important ones are the 3Rs (Replacement, Reduction, and Refinement). Moreover, the Committee for the Purpose of Control and Supervision of Experiments on Animals (CPCSEA) in India has added rehabilitation as the 4th R in regulation guidelines. In India, rehabilitation of the animal models used in a research is controlled as a national policy by CPCSEA. In fact, rehabilitation guidelines necessitate considering postoperative care of animals as a part of study process [30] (◘ Fig. 5.1).

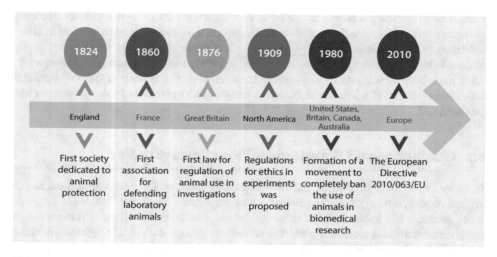

Fig. 5.1 A snapshot of guideline development for preclinical study. Since the nineteenth century, ethics in experimental studies have been drawn consideration especially in animal use and protection field. It was founded first in England and extended to other countries over time [30, 31]

5.8 Challenges and Future Perspectives

During the last decades, considerable investments have been demonstrated in different areas including molecular biology and biomarker study. However, few progression in animal modeling has been accomplished, although there are existing several debates in this area. Humanized animal models have a fundamental role in developing more predictive studies which have remarkable importance in the improvement of other science areas. Hence, animal modeling needs more attention. Consideration of the involving factors which can influence the quality of animal studies seems to be helpful including more and more possible quality improvement of animal models with designing more proper studies according to standard protocols and guidelines. Another improving factor is making correct decisions through using animal models and, finally, trying to develop advances on non-animal models for bridging to the clinic. Some of the potential alternatives for animal models have been introduced by investigators including human-computer models, human organs on chips, and experimental clinical trials. Human subjects, for example, are used for drug effect testing in clinical trials without a large-scale animal testing base. After the number of failures in clinical experiment, clinical protocols will be developed. Accordingly, human models represent special translational importance, but they cannot be a perfect replacement for animal models. Ideally, experimental clinical trials could cause further enhancements in translational studies when it is accompanied to properly design animal studies [28]. In this regard, an animal study is a fundamental step in biomedical researches, and improvement in the design and execution of animal experiments has a particular value.

┌─ Take-Home Messages ───

- Animal models provide improvements in biomedicine, bedside, and benchside.
- Developing an appropriate animal model and critical validation of the provided model lead to minimizing unpredicted adverse effects in clinical trial phases.
- Clinical trial results can be useful for improvement of preclinical studies and vice versa.
- Several animal species have been used for modeling different types of human diseases.
- Animal models can be chosen according to their advantages, limitations, and the purpose of the study.
- A valid animal model represents similar pathophysiologic conditions with humans.
- The validity of animal studies can be improved by following relevant guidelines and standards to achieve more proper study design.

References

1. Ritskes-Hoitinga M, Leenaars M, Avey M, Rovers M, Scholten R. Systematic reviews of preclinical animal studies can make significant contributions to health care and more transparent translational medicine. Cochrane Database Syst Rev. 2014;(3):ED000078.
2. Bailoo JD, Reichlin TS, Würbel H. Refinement of experimental design and conduct in laboratory animal research. ILAR J. 2014;55(3):383–91.
3. Rhrissorrakrai K, Belcastro V, Bilal E, Norel R, Poussin C, Mathis C, et al. Understanding the limits of animal models as predictors of human biology: lessons learned from the sbv IMPROVER species translation challenge. Bioinformatics. 2014;31(4):471–83.
4. Contopoulos-Ioannidis DG, Ntzani E, Ioannidis J. Translation of highly promising basic science research into clinical applications. Am J Med. 2003;114(6):477–84.
5. Goodarzi P, Aghayan HR, Larijani B, Soleimani M, Dehpour A-R, Sahebjam M, et al. Stem cell-based approach for the treatment of Parkinson's disease. Med J Islam Repub Iran. 2015;29:168.
6. Goodarzi P, Payab M, Alavi-Moghadam S, Larijani B, Rahim F, Bana N, et al. Development and validation of Alzheimer's disease animal model for the purpose of regenerative medicine. Cell Tissue Bank. 2019;20(2):141–51.
7. Aghayan HR, Soleimani M, Goodarzi P, Norouzi-Javidan A, Emami-Razavi SH, Larijani B, et al. Magnetic resonance imaging of transplanted stem cell fate in stroke. Journal of research in medical sciences: the official journal of Isfahan University of Medical. Sciences. 2014;19(5):465.
8. Karp NA. Reproducible preclinical research—is embracing variability the answer? PLoS Biol. 2018;16(3):e2005413.
9. Tieu K. A guide to neurotoxic animal models of Parkinson's disease. Cold Spring Harb Perspect Med. 2011;1(1):a009316.
10. Kim YJ, Park HJ, Lee G, Bang OY, Ahn YH, Joe E, et al. Neuroprotective effects of human mesenchymal stem cells on dopaminergic neurons through anti-inflammatory action. Glia. 2009;57(1):13–23.
11. Conn PM. Sourcebook of models for biomedical research. Totowa, New Jersey: Springer Science & Business Media; 2008.
12. Hasenfuss G. Animal models of human cardiovascular disease, heart failure and hypertrophy. Cardiovasc Res. 1998;39(1):60–76.
13. Swanborg RH. Animal models of human disease experimental autoimmune encephalomyelitis in rodents as a model for human demyelinating disease. Clin Immunol Immunopathol. 1995;77(1):4–13.
14. Etuk E. Animals models for studying diabetes mellitus. Agric Biol JN Am. 2010;1(2):130–4.

15. Snyder BR, Chan AW. Progress in developing transgenic monkey model for Huntington's disease. J Neural Transm. 2018;125(3):401–17.
16. De Los Angeles A, Hyun I, Latham SR, Elsworth JD, Redmond DE. Human-Monkey Chimeras for modeling human disease: opportunities and challenges: Chimera Research: Springer; 2019. p. 221–31.
17. Cyranoski, D. Marmoset model takes centre stage. Nature 459, 492 (2009) https://doi.org/10.1038/459492a.
18. Kishi N, Sato K, Sasaki E, Okano H. Common marmoset as a new model animal for neuroscience research and genome editing technology. Develop Growth Differ. 2014;56(1):53–62.
19. Narfström K, Holland Deckman K, Menotti-Raymond M. The domestic cat as a large animal model for characterization of disease and therapeutic intervention in hereditary retinal blindness. J Ophthalmol. 2011;2011:906943.
20. Vandamme TF. Use of rodents as models of human diseases. J Pharm Bioallied Sci. 2014;6(1):2.
21. Shearin AL, Ostrander EA. Leading the way: canine models of genomics and disease. Dis Model Mech. 2010;3(1–2):27–34.
22. Hytönen MK, Lohi H. Canine models of human rare disorders. Rare Dis. 2016;4(1):e1006037.
23. Moran CJ, Ramesh A, Brama PA, O'Byrne JM, O'Brien FJ, Levingstone TJ. The benefits and limitations of animal models for translational research in cartilage repair. J Exp Orthop. 2016;3(1):1.
24. Hodavance MS, Ralston SL, Pelczer I. Beyond blood sugar: the potential of NMR-based metabonomics for type 2 human diabetes, and the horse as a possible model. Anal Bioanal Chem. 2007;387(2):533–7.
25. Varga OE, Hansen AK, Sandøe P, Olsson IAS. Validating animal models for preclinical research: a scientific and ethical discussion. Altern Lab Anim. 2010;38(3):245–8.
26. Kirk AD. Crossing the bridge: large animal models in translational transplantation research. Immunol Rev. 2003;196(1):176–96.
27. Xiong Y, Yu J. Modeling Parkinson's disease in Drosophila: what have we learned for dominant traits? Front Neurol. 2018;9:228.
28. Lazic SE, Essioux L. Improving basic and translational science by accounting for litter-to-litter variation in animal models. BMC Neurosci. 2013;14(1):37.
29. Berge OG. Predictive validity of behavioural animal models for chronic pain. Br J Pharmacol. 2011;164(4):1195–206.
30. Pasupuleti MK, Molahally SS, Salwaji S. Ethical guidelines, animal profile, various animal models used in periodontal research with alternatives and future perspectives. J Indian Soc Periodontol. 2016;20(4):360.
31. Miziara ID, de Matos Magalhães AT, Santos MD, Gomes ÉF, de Oliveira RA. Research ethics in animal models. Braz J Otorhinolaryngol. 2012;78(2):128–31.

Further Reading

Books

Hau J, Schapiro SJ, Van Hoosier GL Jr. Handbook of laboratory animal science: animal models, vol. III.
Conn PM, editor. Animal models for the study of human disease: Academic Press; 2017.
Okechukwu IB. Introductory chapter: animal models for human diseases, a major contributor to modern medicine. In: Experimental animal models of human diseases-an effective therapeutic strategy: IntechOpen; 2018.
Conn PM, editor. Sourcebook of models for biomedical research: Springer Science & Business Media; 2008.
Hau J. Animal models for human diseases. In: Sourcebook of models for biomedical research: Humana Press; 2008. p. 3–8.

Principles of Good Manufacturing Practice

Masoumeh Sarvari, Sepideh Alavi-Moghadam, Bagher Larijani (iD),
Ilia Rezazadeh, and Babak Arjmand (iD)

© Springer Nature Switzerland AG 2020
B. Arjmand et al. (eds.), *Biomedical Product Development: Bench to Bedside*,
Learning Materials in Biosciences, https://doi.org/10.1007/978-3-030-35626-2_6

6.1   What Will You Learn in This Chapter?

In the middle ages, craftsmen were responsible for manufacturing and inspecting their products, and the quality was considered as their own honor. Industrial revolution altered the concept of quality drastically. Rise of inspections and separation of quality departments were the main outcomes of this alteration. On the other hand, statistical methods were the unprecedented approach to control the products variabilities and quality. In addition to statistical methods, "quality control" and "quality management" as novel concepts were the other achievements in the late nineteenth century. Actually, quality has been considered as an organizational idea, and in this organization all participants are responsible. Nowadays, with regard to ever developing and competitive atmosphere, the importance of quality policies and objectives is fully recognized, and also quality management system has been prominent everywhere from research to business and in every aspect of life. This chapter discusses the importance of quality and its performance in our lives and provides a brief description and background of GMP and some basic principles, guidelines, and modules of GMP.

6.2   Rationale and Importance

Recently, the importance of quality is globally accepted because the developments of each country depend on the applied standards and managements. Consequently, quality has become a burning issue for a growing number of countries. Therefore, products and services are no longer considered sufficient if they are not accompanied by quality and supported by adequate quality frameworks and systems [1, 2]. The quality concept is propounded with special sensitivity and greater attention to safety issues in biomedicine. The main purpose of good manufacturing practices (GMP) is to decrease the probable risks that may affect the end products/services. Some of the serious risks in healthcare systems include unintentional contamination, insufficient or too much active ingredients, mislabeling, etc., that can lead to ineffective treatment, adverse effects, and even death. In this regard, due to some reasons such as growing need for qualified human cell and tissue, increasing risk of contamination, and international harmonization, the quality management system (QMS) has found a special position in biomedical science. QMS is a set of all organizational quality principles applying as a guidance in establishments and quality assurance/quality control (QA/QC) procedures. QMS is a cumulative activity to produce and maintain a product or service with desired quality requirements against minimum costs. Altogether, QC, GMP, and QA as interrelationship concepts are subset of each other (Fig. 6.1). QC as the basic level of quality management is a set of procedures for checking or testing to verify and certify the end product/service accordance with the required quality criteria. QC is a subset of GMP with laboratory-based procedure. GMP as a guidance ensures that the required quality will be continued consistently. QA as the proactive organized arrangements is wider than QC and GMP. Further, QA is applied to the developing products/services to ensure that the output will meet the required quality specifications. In other words, QA as an organization-based process can plan and manage the required guidelines and standards to ensure quality.

◘ Fig. 6.1 Quality relationship. QM, QA, GMP, and QC are interrelated concepts. The main target of these concepts is to attain total quality

6.3   Definition

GMP is described as "a part of quality assurance which ensures that products are consistently produced and controlled to the quality standards appropriate to their intended use and as required by the marketing authorization and products specifications." On the other hand, "GMP is that part of quality management which ensures that products are consistently produced and controlled to the quality standards appropriate to their intended use and as required by the marketing authorization, clinical trial authorization or product specification" is another definition of GMP by the European Union (EU) [3]. GMP has various definitions all over the world, but all of them have similar principles, and their ultimate objectives are the same. In other words, GMP as a main objective of health systems includes a set of regulations, codes, and guidelines that describe the methods, equipment, and facilities required to manufacture and control biomedical products based on the appropriate standards. It covers variable areas such as pharmaceutical, biological, and cosmetic products, medical equipment, packaging and labeling, laboratory controls, etc. Safety, purity, identity, strength, and quality of a product can be ensured through the optimal commitment of GMP [4]. Implementation of GMP strictly depends on a rational inspection order. However, processes of GMP are strongly involved in leadership, collaboration, and consistency. Testing the quality of each batches of products during the manufacturing process instead of testing the entire products together at the same time in the batch release step is one of the basic tenets of GMP. In other words, the intended quality cannot be attained only through the final testing and detecting the errors. Hence, controlling and designing the quality into process and preventing the errors during the process will be more efficient and cost-effective [5, 6].

In spite of differences between GMP and cGMP, due to their similar final aims, sometimes they are used instead of each other. The word "current" in the cGMP emphasizes applying the most current and up-to-date techniques and methods, as well as expectations are dynamic. Therefore, this leads to making cGMP a more effective and efficient approach. On the other hand, applying novel technologies in cGMP makes it more reliable in QA than GMP. Given the fact that novel available technologies are more expensive than the

Table 6.1 Comparison between characteristics of GMP* and cGMP** as two main elements of quality management systems

Characteristics	GMP	cGMP
Applied to more than 100 countries	+	−
Application of novel and current technologies	+	++
Cost-effective	++	+
High-quality assurance value	+	++
Much broader	++	+

*Good manufacturing practices
**Current good manufacturing practices

old ones, implementation of cGMP is more expensive than GMP. Additionally, GMP is a broad concept and covers a wide range of situations and areas of science and business. However, cGMP is only limited to manufacturing processes. Finally, GMP is applied by more than 100 countries although only a few number of them adhere to cGMP principles. Altogether, in healthcare systems, the cGMP is more competent than the GMP in achieving the best outcomes (▪ Table 6.1).

6.4 From Nineteenth Century to Now: History of GMP

In the early 1900s, home remedies, ointments, and "miracle elixirs" were applied to treat any discomforts. In the late nineteenth century, the production of vaccines without any regulatory controls was begun in a large amount. After in 1901, antitoxin derived from blood serum of horses was regularly used to treat diphtheria patients. Given the fact that these antitoxins were prepared locally and there was no uniform control, in St. Louis, Missouri, 13 children died from tetanus after treatment with the antitoxin from an infected horse named Jim [7]. In the same year, similar catastrophe happened with contaminated smallpox vaccine. In reaction to such incidents, the importance of health and high-quality raw materials was propounded officially for the first time [8]. Until 1902, there was no regulation to control these medications or vaccines, and the mentioned undesired events led to enact the "Biologics Control Act of 1902" [5]. It was the first act for controlling the biological products. Following to a published data about the contaminated meats in Chicago, the Pure Food and Drug Act in 1906 forbids selling unsanitary meat and fake labeling for the first time [9, 10]. Further, in 1941, a company's sulfathiazole tablets were adulterated with a kind of sedative called phenobarbital. Due to this negligence, 300 people died or got injured. According to this tragedy, FDA decided to reconsider and alter the QC and manufacturing requirements basically. Following these alterations, GMP was born in 1941. Then in the 1960s, thousands of children in the Europe were born with severe limb anomalies because their mothers had used thalidomide as a morning sickness pill. This phenomenon led to strengthen the FDA's regulations such as compulsory animal testing before human trial and obtaining an inform consent from trial participants. Also, in accordance with the new amendment in addition to the safety, the effectiveness of the final product must be proved before marketing [5, 11]. The Laboratory of Hygiene of the Marine Health Service in Washington, D.C., was renamed to the Hygienic Laboratory of the Public Health and Marine Hospital Service in 1902. Subsequently, in 1948, its name was changed to "National Institutes of Health" (NIH) which covered a set of institutes related to biomedical researches. In 1972, the regulations of biological products were transferred to FDA. Eventually in 2010, the FDA Center for Biologics Evaluation and Research (CBER) started to control the safety of biological products as its main responsibility.

6.5 Good Manufacturing Practice Principles

GMP guidance is aimed to ensure that the outcome is safe for human consumption or use. All established GMP requirements follow a number of basic principles that are similar in nature. WHO-GMP guidelines have considered all requirements of different GMP texts and requirements especially in international trade arena in developing countries. The most compliant principles of GMP include (1) designing and constructing the facilities and equipment properly and identifying the responsibilities; (2) following written procedures and instructions; (3) documenting work; (4) validating the processes and evaluating the staff performances; (5) monitoring and regular inspections of facilities and equipment which prevent the accidents; (6) writing step-by-step operating procedures and instructions; (7) designing, developing, and demonstrating job competence; (8) protecting against contamination and promoting of the workplace quality and safety; (9) controlling the components and product-related processes and ensuring the quality of materials; and (10) conducting planned and periodic audit checklists that help to recognize the errors immediately and refine the noncompliant processes. In the road of globalization, more than 100 countries have accepted and followed the GMP guidelines as general standards, and they have mandated their manufacturers to make their own guidelines according to GMP principles. Given this universal acceptance, GMPs have become a principle in importing and exporting healthcare products/services which can facilitate the product/service presentation in a similar setting worldwide. Furthermore, an effective GMP can assist biopharmaceutical companies to get more qualified production, more profit, and less wastes.

6.6 Good Manufacturing Practice Regulations and Guidelines

GMP covers all aspects of production from raw material testing, premises and equipment, manufacturing control, personal hygiene of staff, laboratory controls, quality control department, packaging and labeling, end product testing, sale, records of medical cases, stability, and sterility of products. Each step in the production should have a precise written guidance to explain the procedures of the step in more details. Additionally, a general system is seriously required to control that the early determined principles are followed consistently at each step. There are several versions, amendments, and extensions of GMP guidelines. The WHO's version of GMP guidelines is less intransigent than European and

US ones, and it is easily applied by developing countries. In addition to WHO, considerations to other local and domestic issues are considered by regulators. On the other hand, harmonization is an important tenet which is necessary for applying the endorsed criteria and achieved through period conferences. Actually, GMP guidelines are a set of general principles covering the entire manufacturing and quality control processes. They are not comprehensive instructions of the process, as each system is responsible to arrange its detailed programs corresponded to established guidance to perform both business and regulatory requirements together.

6.7 Challenges and Future Perspective

Primarily, GMPs should be usable and feasible, possess the social acceptability, and describe the basis of risk adjustments. The best way for achieving these concepts is a proper measuring of customers' demands and effective workforce. Collaboration with agencies is a way to decrease the limitations of implementation of GMP. Subsequently, GMP regulations are enforced by different authorities all over the world, for example, FDA in the USA, MHRA in the UK, Therapeutic Goods Administration in Australia, Ministry of Health in India, and Iran Food and Drug Administration (IFDA) in Iran. Unfortunately, there are still impressive number of underdeveloped countries that are not in compliance with GMPs. Additionally, the mentioned and other authorities in other countries are faced with lack of trained staff. Sometimes staff do not have enough expertise and experiences. Also, generally there is no direct WHO supervision on these agencies and staff. Additionally, there is a prominent cultural misinterpretation of the GMP guidelines which needs to be overcome by providing national and regional guidelines and standards focusing on principles of international ones.

Generally, GMP has become an essential element in local and universal marketing. However, the compliance with GMP has not been commonly adopted in developing societies. Therefore, the governments are endured to invest heavily for upgrading the related standards, equipment, facilities, and man powers. From another point of view, these large investments raise the costs, and coping with this issue has become a troublesome domestic marketing of developing countries. Also, small size industries with less developed technologies make the competition with their counterparts difficult. "Market perspective" in developing countries is another difficulty that the governments are facing. Consequently, the manufacturers repine that they can profit without GMPs, and GMP execution has not expected profit for them. After all, strict GMP implementation for these countries makes an obstacle to develop [12]. One of the important tenets of GMP is to apply the novel technologies for enhancing its efficiency. Using current and updated data subsequently has been shown own benefits in the cGMP setting. However, application of novel technologies has its own limitations such as long-term finance, required facilities, adaptation of companies and related authorizations with them, etc. In summary, implementation of GMPs focusing on all their aspects and principles and trying to achieve a globalized setting is a major obstacle worldwide.

Take-Home Message

- GMP is a set of quality assurance practices which are aimed to guarantee the safety and good quality of products.
- GMP guidelines are used across 100 countries, for example, Australia, Europe, China, the Philippines, and Iran, while a few number of the countries are complied with the cGMP guidelines in comparison with GMP.
- GMP is enforced by different authorities all over the world. For example, FDA is responsible for the enforcement of GMP in the USA.
- There are different versions of GMP, but the WHO's version is the general reference, and most of the countries develop their own guidelines based on WHO-GMP following the similar basic principles.
- Every GMP guideline follows the 10 basic principles of WHO version.

References

1. Naor M, Goldstein SM, Linderman KW, Schroeder RG. The role of culture as driver of quality management and performance: infrastructure versus core quality practices. Decis Sci. 2008;39(4):671–702.
2. Young JH, editor. Sulfanilamide and diethylene glycol. ACS symposium series; 1983: AMER Chemical Soc 1155 16TH ST, NW, Washington, DC 20036.
3. Organization WH. WHO Expert Committee on specifications for pharmaceutical preparations: fiftieth report: World Health Organization; 2016.
4. Sensebé L, Gadelorge M, Fleury-Cappellesso S. Production of mesenchymal stromal/stem cells according to good manufacturing practices: a review. Stem Cell Res Ther. 2013;4(3):66.
5. Shadle PJ. Overview of GMPs. Biopharm International. 2004:8–14.
6. Sensebé L, Bourin P, Tarte K. Good manufacturing practices production of mesenchymal stem/stromal cells. Hum Gene Ther. 2010;22(1):19–26.
7. Baylor NW. Role of the national regulatory authority for vaccines. Int J Health Gov. 2017;22(3):128–37.
8. Patel K, Chotai N. Pharmaceutical GMP: past, present, and future–a review. Die Pharmazie Int J Pharm Sci. 2008;63(4):251–5.
9. Immel BK. A brief history of the GMPs for pharmaceuticals. Pharm Technol. 2001;25(7):44–53.
10. Blackwell J. 1906: Rumble over 'The Jungle'. The Capitol Century. 1998;17.
11. Lee PR, Herzstein J. International drug regulation. Annu Rev Public Health. 1986;7(1):217–35.
12. Harper I, Brhlikova P, Pollock A. Good manufacturing practice in the pharmaceutical industry–working paper 3, prepared for workshop on 'Tracing Pharmaceuticals in South Asia', 2–3 July. University of Edinburgh, 2007.

Further Reading

Books

Karmacharya JB. Good Manufacturing Practices (GMP) for medicinal products, 2012.
Bunn GP. Good manufacturing practices for pharmaceuticals. 7th ed; 2019.
Cooper BN. The GMP handbook: a guide to quality and compliance, 2017.
Cooper BN. Good manufacturing practices for pharmaceuticals: GMP in Practice, 2017.
Oechslein Ch. GMP fundamentals – a step-by-step guide for good manufacturing practice, 2015.
Arjmand B. Perinatal tissue-derived stem cells alternative sources of fetal stem cells. 2016.
Nally JD. Good manufacturing practices for pharmaceuticals, 2016.

Online Resources

FDA. 21 CFR 210, 211, 800. Available at: http://www.fda.gov/.
FDA. Milestones in US food and drug law history. Available at: http://www.fda.gov/opacom/background-
    ers/miles.html.
ICH. www.ich.org.
WHO. www.who.int.
EU/EMEA. www.emea.europa.eu.

6

The Importance of Cleanroom Facility in Manufacturing Biomedical Products

Sepideh Alavi-Moghadam, Masoumeh Sarvari, Parisa Goodarzi, and Hamid Reza Aghayan

© Springer Nature Switzerland AG 2020
B. Arjmand et al. (eds.), *Biomedical Product Development: Bench to Bedside*,
Learning Materials in Biosciences, https://doi.org/10.1007/978-3-030-35626-2_7

7.1   What You Will Learn in This Chapter?

This chapter is trying to describe:
- The fundamental role of cleanroom facility in the manufacturing of biomedical products
- The main purposes of the cleanroom design
- The definition and features of cleanroom facility based on the international standards
- The sources of contamination in cleanroom facility
- The practical approaches for contamination control in cleanroom facility
- The cleanroom classification
- The principles of cleanroom validation and monitoring

7.2   Rationale and Importance

Since the manufacturing process of biomedical products (for clinical applications) requires a number of safety considerations, their production must be performed based on the principles of certain standards [1–4]. In this context, the use of clean and non-contaminated spaces and the application of sterile raw materials and aseptic techniques can play a crucial role. The cleanroom as a controlled and clean area should be maintained according to some specific parameters including the temperature, humidity, air pressure, and particles [5, 6]. Accordingly, the design and installation of a cleanroom facility are very complicated and should be done by the expert and qualified group (highly experienced designers and also GMP specialists) [6, 7]. Moreover, cleanrooms have a wide range of applications based on the aim and scope of each institute [5], and there are different cleanroom classification methods [6, 8]. On the other hand, to ensure that the working environment and equipment meet the regulatory requirements, cleanroom validation should be performed [9]. In general, clean and controlled area such as a cleanroom is needed to produce safe and effective biomedical products.

7.3   The Definition of Cleanroom

A cleanroom is defined as a facility that is employed for a particular manufacturing of products or scientific investigations [1, 5]. In accordance with the definition of the US Federal Standard 209 (FS 209), a room which has a certain amount of airborne particles is considered as a cleanroom [10]. On the other hand, ISO (International Organization for Standardization) 14644-1 standard has introduced a cleanroom as an area in which the air-suspended particles and the temperature, humidity, and air pressure are controlled. Additionally, according to ISO 14644-1 standard, the cleanroom should be designed and used in a manner that the number of suspended particles in the air should not exceed the maximum predefined level of each class [11].

7.4   Purpose and Strategy of a Cleanroom Design for Manufacturing of Biomedical Products

The objectives can be evaluated as follows [5, 6, 9, 12]:
- Controlling contaminants caused by the production process
- Controlling the cross contamination among two or more procedures and products

- Controlling contaminants caused by personnel operations
- Optimizing the arrangement and rooms connection to control and manage personnel movements
- Providing personnel protection
- Providing a safe and comfortable working atmosphere
- Precise designing and configuration for controlling the material entry and exit during the production process
- Controlling any potential hazards arising from product and procedures
- Efficient maintenance and monitoring of the cleanroom

Hence, in order to meet the mentioned purposes, preparing the process flow diagrams, analyzing the production process, and introducing the specific activities and requirements for environmental quality should be considered.

7.5 Contamination Sources

The process or activity which leads to soiled, stained, and infected materials or surfaces is called contamination [13, 14]. Therein, some of the important contamination sources of cleanroom environment are including *people* (e.g., skin flakes, cosmetics and perfume, clothing, and hair), *facilities* (e.g., walls, floors, ceilings, paint and coatings, and air conditioning debris), *fluids* (e.g. bacteria, organics, moisture, cleaning chemicals, and deionized water), *equipment and supplies* (e.g., wear particles, outgasses, brooms, mops, and dusters), and *product generated* (e.g., silicon chips, quartz flakes, and aluminum particles). Control and prevention of contamination entering into the cleanroom is an important issue that needs to be managed [5, 15].

7.6 Contamination Control

Some of the basic and essential principles of contamination control in the cleanroom are as follows:

- **Proper design of the cleanroom**

The cleanroom architecture and design, the connection between different parts, appropriate process flows, and the methods of optimizing performance and reducing power consumption are very important [6, 9, 16]. Herein, the control of temperature, humidity, air pressure differential and flows between the rooms (to create sustainable conditions for manufacturing process along with prevention of contamination) and optimizing personnel, equipment, materials, waste, and products flows based on the international regulations and efficient operation of the cleanroom should be considered. ▶ Box 7.1 demonstrates some requirements of the cleanroom internal structure [17–20]. Generally, in addition to matters related to electrical, mechanical, behavioral, etc., the architectural factors (including dimensional and geometric features of the space, access and entrance control, and materials of structural components) are also important in controlling the contamination of the cleanroom [1, 15, 21].

Box 7.1   Some Requirements of the Cleanroom Internal Structure
- Walls, floors, and ceiling should be washable, smoothed, non-porous and without any cracks and cavities.
- Ceiling-wall and wall-floor junction site should be concave.
- Ceilings and equipment should have minimum contamination accumulation.
- Doors should be easily washable and should not absorb moisture.
- Lamps must be covered in a way that can be washed, and their light intensity should allow for normal operation.
- Air pressure of the room with a higher degree of cleanliness should be more positive than others.
- To avoid cross contamination and connection between the cleanroom air and the outside air, airlocks should be used in the path of personnel and material passage.
- To decrease the particle contamination, specific enclosed antechamber named as air shower (◘ Fig. 7.1) which employ high-pressure air filters (HEPA or ULPA) can be used in entrance of cleanrooms.

7

◘ **Fig. 7.1** Air shower. Air shower is an enclosed antechamber which is installed in the entrance of cleanrooms (usually between change areas and the cleanroom). It removes particles from personnel or object surfaces

■   **The use of the heating, ventilation, and air conditioning (HVAC) system**

One of the most important parts of cleanroom is HVAC system which is considered the heart of this facility. In general, some factors such as the amount of air-suspended particles are influenced by air pressure, air flow, air changes per hour, temperature, and humidity (indoor air quality) are controlled by the HVAC system [22, 23].

■   **The use of an appropriate cleanroom suit**

Personnel as one of the main sources of contamination must follow some standard rules to reduce contamination, such as the use of appropriate cleanroom suit. Appropriate cleanroom suit includes hood, safety glasses, face mask, main garment (coverall dress), sterile gloves, and boots [1, 5] (■ Fig. 7.2).

■   **Cleaning and washing the cleanroom**

Constantly, cleanroom cleaning and full contamination control should be done using cost-effective materials during a quick and easy process. In this respect, for all surfaces, floors, walls, and benches, applying the cleanroom mops with DI water (deionized water) and appropriate disinfectant is suggested [1, 24].

■ **Fig. 7.2**   Cleanroom suit. Personnel as one of the main sources of contamination must be trained and use the appropriate cleanroom suit. Appropriate cleanroom suit includes hood, safety glasses, face mask, main garment, sterile gloves, and boots

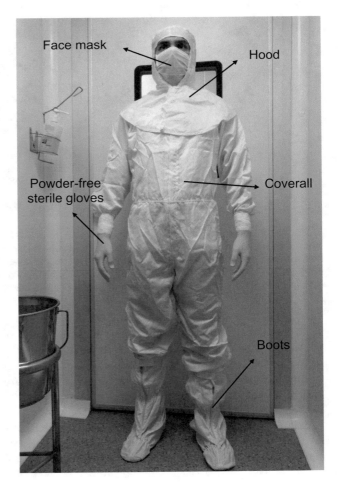

7

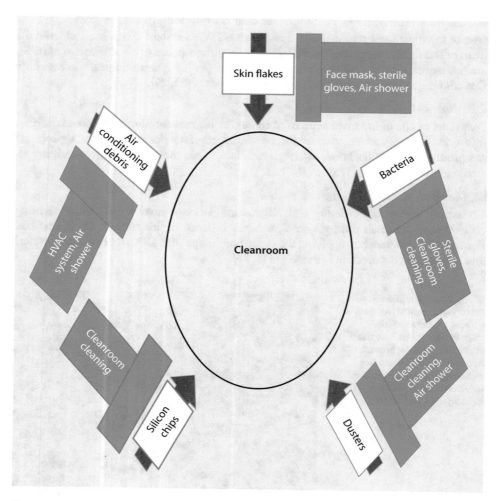

◘ Fig. 7.3 Possible cleanroom contaminations and their control. Some of the important contamination sources of cleanroom are people (e.g., skin flakes,), facilities (e.g., air conditioning debris), fluids (e.g., bacteria), equipment and supplies (e.g., dusters), and product generated (e.g., silicon chips) which can be controlled by some of the basic and essential contamination control principles such as using HVAC system, positive air pressure, cleanroom cleaning, appropriate cleanroom suit, etc

■ **Regular cleanroom monitoring and validation**

To ensure the correct cleanroom design, installation, and performance, all of the related procedures should be validated precisely. In accordance with ISO 14644-1, cleanroom validation is performed through three steps: (1) as built (a complete room with all services connecting and functioning without any equipment, furnitures, material, or personnel present), (2) at rest (all equipment is installed and operated, but no personnel present), and (3) in operational (all equipment is installed and in operation; personnel are also present and are working) [1, 25]. On the other hand, viable and non-viable airborne particles, air velocity, humidity, and temperature should be monitored regularly using settled plates, air sampler and particle counter, anemometer, and temperature/humidity controller, respectively [26].

Some of the possible cleanroom contaminations and their control are shown in ◘ Fig. 7.3.

7.7 Cleanroom Classification

Cleanroom classification methods have been developed based on amount and size of airborne particles per volume of the air [5]. In summary, there are different cleanroom classifications based on various standards (◘ Table 7.1). For instance, FS 209 has classified cleanrooms into class 100 000 (the dirtiest class), 10 000, 1000, 100, 10, and 1 (the cleanest class). The ISO classification system has introduced the cleanroom classes as ISO 1 – ISO 9 (in which the cleanest class is ISO 1 and the dirtiest class is ISO 9). The GMP (good manufacturing practice) guideline has classified cleanrooms into four different grades (◘ Fig. 7.4). In this context, there are four types of clean zones (based on product types and the level of the process that should be protected from contamination). Accordingly, high-risk activities (such as opening and closing bottles containing essential sterile materials for the manufacturing processes) should be done in zone A. Additionally, equipment that provide aseptic environments (e.g., laminar air flow safety cabinet) are considered as zone A. Zone B is an intermediate space (buffer area) for preparing aseptic products, materials, and equipment that are intended to enter grade A. Finally, the procedures with lower risks can be performed in zones C and D [1, 11].

7.8 Cleanroom Air Flow

Cleanrooms can control concentration of the air particles using HEPA or ULPA filters through laminar or turbulent airflow. Accordingly, cleanrooms are divided into three groups [8, 27–29] (◘ Fig. 7.5):
1. Cleanrooms with laminar air flow (unidirectional airflow): Laminar airflow has direction in a constant stream, and cleanrooms with laminar air flow are including cleanrooms with *horizontal* (the direction toward is filtered which is located on walls) and *vertical* airflow (the direction from the ceiling to the floor). Further, cleanrooms with laminar air flow have the highest degree of cleanliness.
2. Cleanrooms without laminar air flow (non-unidirectional airflow): It employs turbulent airflow systems to control air particles.
3. Cleanrooms with mixed airflow: They have both unidirectional and non-unidirectional airflow.

7.9 Challenges and Future Perspective

The most important cleanroom problems and challenges are related to designing, construction, operation, monitoring, and maintenance. Some of them include insufficient designing, poor ventilation, and hardworking maintenance and operation. To overcome these challenges, there are several approaches such as convenient management, personnel training, and using validated SOPs (standard operating procedures). In another side, according to new biomedical product manufacturing requirements, future cleanrooms will focus on profitability and economic efficiency. Herein, the advances in new technologies will have effective roles in order to meet these requirements.

7

☐ **Table 7.1** Cleanroom classification based on EU GMP, FS 209, and ISO [1]

EU GMP classification	FS 209	ISO classes	Maximum particles/m³				
			≥0.1 μm	≥0.2 μm	≥0.3 μm	≥0.5 μm	≥1 μm
		ISO1	10	2.37	1.02	0.35	0.083
		ISO2	100	23.7	10.2	3.5	0.83
	Class 1	ISO3	1000	237	102	35	8.3
	Class 10	ISO4	10,000	2370	1020	352	83
Grade A	Class 100	ISO5	100,000	23,700	10,200	3520	832
	Class 1000	ISO6	1,000,000	237,000	102,000	35,200	8320
Grade B	Class 10000	ISO7	10,000,000	23,700,000	1,020,000	352,000	83,200
Grade C	Class 100,000	ISO8	100,000,000	237,000,000	10,200,000	3,520,000	832,000
Grade D	Room air	ISO9	1,000,000,000	2,370,000,000	102,000,000	35,200,000	8,320,000

EU GMP European Union's good manufacturing practices, *FS 209* Federal Standard 209, *ISO* International Standards Organization

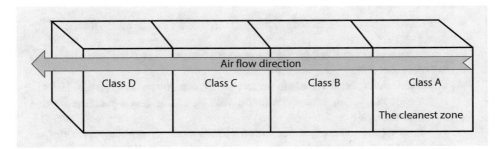

■ **Fig. 7.4** Pressure cascade in cleanroom. Class A is considered as a more stringently controlled area and the cleanest zone with the most positive air pressure. The air flow direction is from class A to class D. High-risk activities should be done in classes A and B [1]

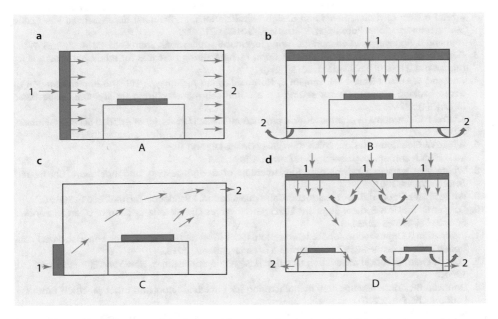

■ **Fig. 7.5** Different patterns of cleanroom airflow. **a** laminar horizontal airflow, **b** laminar vertical airflow, **c** non-unidirectional air flow, and **d** mixed airflow. (1) Airflow entry point. (2) Airflow outlet point

> **Take-Home Messages**
>
> - The manufacturing of biomedical products should be performed in a controlled area.
> - Cleanroom as a controlled area is the proper environment for biomedical product manufacturing.
> - Contamination control in cleanroom can be achieved by some strategies including personnel training, proper designing, validated SOPs, and regular monitoring and maintenance.
> - To ensure cleanroom compliance with the GMP standards, regular monitoring and validation are required.

References

1. Aghayan HR, Arjmand B, Burger SR. GMP facilities for clinical cell therapy product manufacturing: a brief review of requirements and design considerations. In: Perinatal tissue-derived stem cells: Springer International Publishing, Switzerland; 2016. p. 215–27.
2. Arjmand B, Aghayan HR. Cell manufacturing for clinical applications. Stem Cells. 2014;32(9):2557–8.
3. Gouveia BG, Rijo P, Gonçalo TS, Reis CP. Good manufacturing practices for medicinal products for human use. J Pharm Bioallied Sci. 2015;7(2):87.
4. Arjmand B, Emami-Razavi SH, Larijani B, Norouzi-Javidan A, Aghayan HR. The implementation of tissue banking experiences for setting up a cGMP cell manufacturing facility. Cell Tissue Bank. 2012;13(4):587–96.
5. Kitain M. Cleanrooms in pharmaceutical production. Mikkeli University of Applied Sciences, Finland; 2010.
6. Whyte W. Cleanroom design. Chichester: Wiley Online Library; 1999.
7. Mills HD. Cleanroom engineering. Adv Comput. 1993;36:1.
8. Whyte W. Cleanroom technology: fundamentals of design, testing, and operation. Chichester/Toronto: Wiley; 2001.
9. White E. Cleanroom design, construction, and qualification. J Validation Technol. 2009;15(4):30.
10. Cooper D. Towards Federal Standard 209E: partial versus complete inspection of clean air zones. J Environ Sci. 1989;32(3):31–3.
11. International Organization for Standardization. ISO 14644-1:2015, Cleanrooms and associated controlled environments—Part 1: Classification of air cleanliness; 2015.
12. Cole G. Pharmaceutical production facilities: design and applications. New York: Ellis Horwood Ltd; 1990.
13. Donovan RP. Contamination-free manufacturing for semiconductors and other precision products. London: CRC Press; 2001.
14. Ramstorp M. Introduction to contamination control and cleanroom technology. New York: Wiley; 2008.
15. Ljungqvist B, Reinmuller B. Clean room design: minimizing contamination through proper design: CRC Press, United State of America; 1996.
16. McFadden R. A Basic Introduction to Clean Rooms 2008. Available from: http://www.coastwidelabs.com/Technical%20Articles/Cleaning%20the%20Cleanroom.htm.
17. Bhatia B. A basic design guide for clean room applications. PDH Online. 2012;143:1–61.
18. Zhang J. Pharmaceutical cleanroom design. ASHRAE J. 2004;46(9):29–34.
19. Lee JS, Park JK, Jun JK, Choi SB. Air-shower system for a clean room. Google Patents; 1997.
20. Mattson B. Air shower with directed air flow. Google Patents; 1990.
21. Gee A. Design of a new GMP facility-lessons learned. In: Cell therapy. Boston: Springer; 2009. p. 79–84.
22. McQuiston FC, Parker JD. Heating, ventilating, and air conditioning: analysis and design. Chichester: Wiley; 1982.
23. Schneider RK. Designing clean room HVAC systems. ASHRAE J. 2001;43(8):39.
24. Paley WR, Paley SJ, Cooper DW, Russo PB, Sayre JC, Siegerman HD, et al. Cleaning method. Google Patents; 1998.

7

25. Caselli-Fernández LM, Terkola R. Clean room environment, personnel, quality assurance and their monitoring. Eur J Hosp Pharm Pract. 2006;12:29–34.
26. Lee J-S, Lim K-H, An Y-H, Choi J-H. Monitoring system and monitoring method for a clean room regulating system. Google Patents; 2001.
27. Xu Z, Zhou B. Fundamentals of air cleaning technology and its application in cleanrooms. Berlin: Springer; 2014.
28. Schicht HH. Clean room technology: the concept of total environmental control for advanced industries. Vacuum. 1985;35(10):485–91.
29. International Organization for Standardization. ISO 14644-4:2001, Cleanrooms and associated controlled environments – Part 4: Design, construction and start-up; 2001.

Further Reading

Books

Gee A. Cell therapy: cGMP facilities and manufacturing. New York: Springer Science & Business Media; 2009.
Ramstorp M. Introduction to contamination control and cleanroom technology. New York: Wiley; 2008.

Articles

Giancola R, Bonfini T, Iacone A. Cell therapy: cGMP facilities and manufacturing. Muscles Ligaments Tendons J. 2012;2(3):243.
Lindblad R. Regulation of cell product manufacturing and delivery: a United States perspective. In: Cell therapy. Boston: Springer; 2009. p. 3–25.

Safety Concerns and Requirement of Cell-Based Products for Clinical Application

Fereshteh Mohamadi-Jahani, Mina Abedi, Maryam Arabi, Akram Tayanloo-Beik (iD)*, and Bagher Larijani* (iD)

© Springer Nature Switzerland AG 2020
B. Arjmand et al. (eds.), *Biomedical Product Development: Bench to Bedside*,
Learning Materials in Biosciences, https://doi.org/10.1007/978-3-030-35626-2_8

8.1   What You Will Learn in This Chapter

Human stem cell-based therapies present alternative promising approach to various diseases in recent years. One of the important factors in this regard is safety. Challenges of safety consideration and efficacy of human cell-based products need to be further addressed. Safety concerns depend on many risk factors including intrinsic, extrinsic, and associated with clinical characteristics. This chapter explains importance of safety. Thereafter, several risk factors will be described.

Altogether cellular therapy accounts as a hopeful therapeutic strategy for numerous diseases, complete investigation has been demonstrated.

8.2   Rationale and Importance

The aim of cellular therapy and regenerative medicine is repairing, restoring, or regenerating damaged cells and tissues. This field has grown as an effective treatment for acute and chronic diseases in recent years [1]. Accordingly, one of the most important subjects in this field is safety [2]. The Food and Drug Administration (FDA) has established appropriate regulatory standards for cell- and tissue-based products. For instance, Code of Federal Regulation for Food and Drugs (21-CFR) is emphasized on safety [3]. Also, good manufacturing practices (GMP) as a component of regulatory structure can provide safety requirements [4]. Finally, safety is one of the most important necessities of clinical grade manufacturing stem cell-based products.

8.3   Classification of Risk Factors

8.3.1   Intrinsic Risk Factors

8.3.1.1   Origin of Cells

Stem cells can be derived from allogeneic or autologous tissues. Both of them have different sources, for instance, adipose tissue, bone marrow, skin, cord blood, peripheral blood, etc. Each of the mentioned origins has totally various safety considerations. In accordance with the above, rejection in allogeneic transplantation is a major risk. Transmission of diseases from the donor to the recipient is also an important concern. On the other hand, underlying disease and aging have a notable impact on the number and functionality of the stem cells in the autologous transplantation. Although autologous cell therapy is introduced as a safe treatment, allogeneic one also is defined as a lower-risk product in the second degree [4–6].

8.3.2   Characteristics of Cells

8.3.2.1   Differentiation Status

The self-renewal and differentiation status of a stem cell play a pivotal role in the clinical outcome. Differentiation potential of stem cells is seriously decreased from embryonic stem cells (ESCs) to adult ones. Further, clinical outcome of dedifferentiation of adult stem

Safety Concerns and Requirement of Cell-Based Products for Clinical...

83

8

cells remains unknown, but mesenchymal stem cell (MSC) differentiation into undesired cell types such as osteocytes and adipocytes is demonstrated in some studies which can theoretically induce cellular alteration and may be an anticipated consequence [5, 7].

8.3.2.2 Tumorigenicity

Tumorigenicity as a major concern should be evaluated before, during, and after stem cell therapy. Tumorigenicity is a native trait in ESCs [8]. Undifferentiated ESCs impose a major risk on cell therapy due to their unregulated cell growth and tumor formation potential. Accordingly, tumorigenicity assay is a requirement in the manufacturing of clinical products [9]. Moreover, ethical and legal limitations can be added to several obstacles of ESC application [10].

8.3.2.3 Proliferation Capacity

Proliferation capacity is required for expansion of stem cells. On the other hand, it has been proven that bone marrow mesenchymal stem cells (BM-MSCs) will be stable genetically after a period of expansion. Moreover, in some cases, cellular senescence happens at late subcultures and results in the change of morphology and immunophenotype and also decrease in proliferation capacity. These changes can induce some undesired risks additionally [11].

8.3.2.4 Lifespan

Most adult stem cells have a limited cell division and self-renewal capacity. Accumulation of genetic, nongenetic, and environmental factors resulted in aging process [12, 13]. Although senescence has adverse effects on stem cells, unlimited cell proliferation results in a prolonged lifespan and malignant transformation. In summary, too long or too short lifespan has its own risks and concerns. Therefore, scientists should balance their advantages with disadvantages [14].

8.3.3 Extrinsic Risk Factors

The extrinsic risk factors are related to lack of donor history, contamination, tumorigenic potential (induced transformation; iPS), conservation (e.g., cryopreservatives), storage and transport conditions (traceability, labelling), and cellular dosage and regimen.

8.3.3.1 Lack of the Donor History

Donor medical history has a fundamental role in formulating donor eligibility and avoiding risks that can be imposed due to the lack of medical information especially in an allogeneic setting [5, 15]. Records or other information received from clinical sources can help scientists and clinicians to control communicable disease transmission from the donors to the recipients. Accordingly, donor eligibility criteria have been developed by relevant austerities such as FDA focusing on donor screening test for viral diseases [16].

8.3.3.2 Contamination

Contamination can occur in the different steps of manufacturing cell-based products. To achieve a reasonable safety in cell therapy, a clean and aseptic environment should be established to avoid any potential contamination. Thereafter, clean room facility has been used in managing particulates to prevent airborne contamination. On the other hand, raw materials (such as cell culture medium) and the stem cell sources can be an origin of contamination. Therefore, providing a controlled area for retrieving tissue sources is necessary in tissue procurement phases [9, 17]. In this context, it has been seriously recommended that the final product be tested for bacteria, viruses, mycoplasma, endotoxin, etc. prior to release [18].

8.3.3.3 Processing and Storage of Human Cell-Based Products

In order to achieve suitable outcomes and avoid therapeutic failure, paying attention to manufacturing procedures of cell products, storage, and preservation seems critical. According to the manufacturing procedures based on particular therapeutic purpose, cell products could undergo different risks. Moreover, the proper cryopreservation and storage need a well-designed procedure to diminish any alterations in cell characteristics such as viability, immunogenicity, and chromosomal aberrations. In other words, cryopreserved products should be protected against probable contamination during thawing and administration. Also, packaging and transporting processes must be suitably established in accordance with relevant standards to prevent contamination. For this purpose, several approved guidelines have been developed to address all of the cell manufacturing steps to assess and manage the potential risk factors. In this concept, implementation of GMP standards can help scientists overcome relevant risk factors from the processing to cryopreservation, packaging, and transportation [19].

8.3.3.4 Coding and Traceability

The advancement of healthcare technologies leads to better results in patient lifesaving and safety. Nowadays, several sharing networks for organs, tissues, and cells have been established all over the world. Following the global distribution, the pivotal importance of a coding and traceability system has been emphasized. Accordingly, using a standard coding system helps scientists with tracing biological products from donor to recipient and vice versa. In this regard, various directives and guidances have been developed to eliminate different challenges associated with transplantation, such as inherent and rejection risk [20]. In this setting, International Society of Blood Transfusion (ISBT) 128, a global protocol based on Code 128 Barcode, has been introduced by the International Council for Commonality in Blood Banking Automation (ICCBBA) which is a universal standard for coding and labeling of blood component and its derivatives. ISBT-128 has been also recommended as an appropriate tool for coding and labeling of cell- and tissue-based products [21]. Accordingly, labels should consist of different sections such as component code and description, donation identification number, collection date, storage conditions, contents or volume, and expiry date and time. In the case of blood components, labels could have modifier text section to specify additional information. Labels can provide significant information about a specific product. They should confirm eligible characteristics of a specific product. Additionally, the World Health Organization (WHO) has two global consultations that provide information about global circulations of human cell and tissue products [20]. The first one is about improving the assessment of transplantation products which can guarantee suitable regulations. The second one elucidates the minimum level of safety and quality which is required for tissue and cell products. From 1991, the WHO consultations have been updated for addressing more issues in this area [22].

Safety Concerns and Requirement of Cell-Based Products for Clinical...

85

8

8.3.3.5 Cellular Dosage and Regimen

Several investigators in the field of stem cell therapy are dealing with a wide range of conditions associated with different stem cell type variants. With regard to the origin of cells, the cellular dose is one of the determining factors in a therapeutic outcome. Accordingly, establishing an appropriate dose for transplantation is critical for cellular therapy. Especially, it is important to consider the association between cell dosages with the weight of patients. In addition, dose regimen is another determining factor which can optimize cell transplantation therapy. It can be determined considering the disease and patient-related factors [23]. Eventually, regarding the different reports, investigators have still few consensuses in this area [24].

8.3.3.6 Tumorigenecity

Human-induced pluripotent stem cells (hiPSCs) were first reported in 2006. Due to prevention of tumor formation, proliferation and differentiation characteristics of iPSCs should be controlled [25]. Additionally, the genomic instability of iPSCs may happen. Therefore, their use for clinical application can impose serious risks [26]. Generally, uncontrolled differentiation and incidence of malignant transformation reveal the importance of safety concerns related to iPSC-derived products in clinical applications [27].

8.3.4 Clinical Characteristics

8.3.4.1 Administration Route

The selection of the appropriate administration route is a crucial step in the development of new therapies which remain controversial due to the diversity in cell type and different doses. Especially, in the case of diseases with less recognized mechanisms, determination of the suitable dosage and administration route will be more challenging. Hereupon, appropriate preclinical and clinical studies which address the different challenges in this regard are necessary [28]. In summary, preclinical studies can help researchers with developing the most suitable rote and dosage according to the study characteristics [28].

8.3.4.2 Immune Response

As a novel field of science, stem cell therapy sheds new light on treatment of different diseases. However, despite the obvious therapeutic potential of stem cells, their application has some risks that need to be balanced with its benefits reasonably. Herein, the immunogenic potential of stem cells is one of the most common risks which can lead to rejection phenomenon. It has been proposed that immunologic reactions are more prevalent in allogeneic cellular therapy. The immune response occurs due to the nonconformity of different antigens in donor or recipient due to the diversity of polymorphisms in allogeneic cases. In this concept, proper risk assessment and managing potential risks can be accomplished through conducting preclinical studies and clinical trials [18]. Moreover, therapeutic use, indication, use of immunosuppressive, and underlying disease are other factors that impose some additional risks (◘ Fig. 8.1).

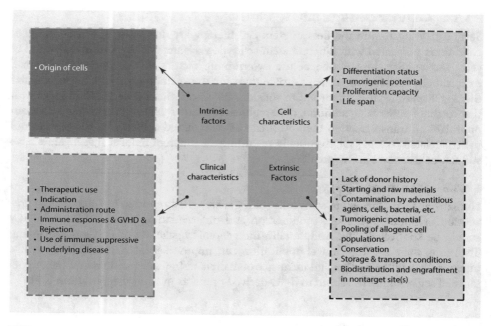

■ **Fig 8.1** Risk factors of cellular therapy. There are several potential risk factors related to cell-based therapy procedures that can be considered as the source of defects. Allogeneic or autologous origin of cells, manufacturing, and handling of cell products and extrinsic factors are potential risks that should be assessed properly [5, 29]

8.4   **Challenges and Future Perspectives**

Regenerative medicine reveals remarkable progression in different fields especially in cell-based therapies. Nowadays, approved cellular products for the clinical application have been dramatically grown and raise several considerations and requirements in the field. By increasing the need for transportation of cell products across the boundaries, these requirements will be more visible. Accordingly, all of cell manufacturing processes should be implemented precisely to ensure safety and efficacy. Cell, tissue, and organ products need to be identified accurately and linked to relevant tests in the donor. Therefore, establishing a coding system according to the global standards leads to the applicable traceability procedure which is globalized by various international authorities such as the WHO. However, despite the development of coding and traceability systems, there are some inevitable risks. Tumorigenicity and immunogenicity are considered as two most important risks which can be minimized using an established traceability procedure for monitoring the origin of calls and materials, cell manufacturing processes, equipment, genetic manipulation, etc. [5]. Also, in autologous and allogeneic transplantation, follow-up of each patient is critical [30]. Finally, while the brilliant progress in stem cell therapy is obvious, implementation of standard systems to achieve more safety and provide high-quality cell products is necessary [27, 31].

┌─ **Take-Home Messages** ──────────────────────────────────

- Achieving an acceptable safety level is essential prior to transplantation.
- Cell processing must be designed to assure safety.
- Cell characteristics, clinical characteristics, extrinsic factors, manufacturing, and handling should be considered in providing biomedical products especially from the safety viewpoint.
- Tumorigenecity is one of the major concerns in cell therapy.
- The implementation of a proper coding system is essential for tracing biological products.
- Cellular dosage and regimen are determining factors in cell transplantation therapy.

└───

References

1. Zheng MH, Pembrey R, Niutta S, Stewart-Richardson P, Farrugia A. Challenges in the evaluation of safety and efficacy of human tissue and cell based products. ANZ J Surg. 2006;76(9):843–9.
2. Rayment EA, Williams DJ. Concise review: mind the gap: challenges in characterizing and quantifying cell- and tissue-based therapies for clinical translation. Stem Cells (Dayton, Ohio). 2010;28(5): 996–1004.
3. Halme DG, Kessler DA. FDA regulation of stem-cell-based therapies. N Engl J Med. 2006;355(16): 1730–5.
4. Burger SR. Current regulatory issues in cell and tissue therapy. Cytotherapy. 2003;5(4):289–98.
5. Herberts CA, Kwa MS, Hermsen HP. Risk factors in the development of stem cell therapy. J Transl Med. 2011;9(1):29.
6. Efimenko AY, Kochegura TN, Akopyan ZA, Parfyonova YV. Autologous stem cell therapy: how aging and chronic diseases affect stem and progenitor cells. Biores Open Access. 2015;4(1):26–38.
7. Breitbach M, Bostani T, Roell W, Xia Y, Dewald O, Nygren JM, et al. Potential risks of bone marrow cell transplantation into infarcted hearts. Blood. 2007;110(4):1362–9.
8. Pera MF, Trounson AO. Human embryonic stem cells: prospects for development. Development. 2004;131(22):5515–25.
9. Yasuda S, Sato Y. Tumorigenicity assessment of human cell-processed therapeutic products. Biologicals. 2015;43(5):416–21.
10. Allum N, Allansdottir A, Gaskell G, Hampel J, Jackson J, Moldovan A, et al. Religion and the public ethics of stem-cell research: attitudes in Europe, Canada and the United States. PLoS One. 2017;12(4):e0176274.
11. Kundrotas G, Gasperskaja E, Slapsyte G, Gudleviciene Z, Krasko J, Stumbryte A, et al. Identity, proliferation capacity, genomic stability and novel senescence markers of mesenchymal stem cells isolated from low volume of human bone marrow. Oncotarget. 2016;7(10):10788.
12. Oh J, Lee YD, Wagers AJ. Stem cell aging: mechanisms, regulators and therapeutic opportunities. Nat Med. 2014;20(8):870–80.
13. Petersen T, Niklason L. Cellular lifespan and regenerative medicine. Biomaterials. 2007;28(26): 3751–6.
14. Kim HJ, Park J-S. Usage of human mesenchymal stem cells in cell-based therapy: advantages and disadvantages. Dev Reprod. 2017;21(1):1.
15. Stroncek DF, England L. Protecting the health and safety of cell and tissue donors. ISBT Sci Ser. 2015;10(Suppl 1):108–14.
16. Food and Drug Administration, HHS. Eligibility determination for donors of human cells, tissues, and cellular and tissue-based products. Final rule. Fed Regist. 2004;69(101):29785.
17. Mizutani M, Samejima H, Terunuma H, Kino-oka M. Experience of contamination during autologous cell manufacturing in cell processing facility under the Japanese Medical Practitioners Act and the Medical Care Act. Regen Ther. 2016;5:25–30.

18. Heslop JA, Hammond TG, Santeramo I, Tort Piella A, Hopp I, Zhou J, et al. Concise review: workshop review: understanding and assessing the risks of stem cell-based therapies. Stem Cells Transl Med. 2015;4(4):389–400.
19. Galvez-Martin P, Hmadcha A, Soria B, Calpena-Campmany AC, Clares-Naveros B. Study of the stability of packaging and storage conditions of human mesenchymal stem cell for intra-arterial clinical application in patient with critical limb ischemia. Eur J Pharm Biopharm. 2014;86(3):459–68.
20. Strong DM, von Versen R. Coding and traceability for products of human origin. Cell Tissue Bank. 2010;11(4):325–7. https://doi.org/10.1007/s10561-010-9223-3.
21. Pruß A. Coding of tissue and cell products. Transfus Med Hemother. 2017;44(6):382.
22. World Health Organization. Second global consultation on regulatory requirements for human cells and tissues for transplantation: towards global harmonization through graduated standards. WHO Report. 2006.
23. Gyurkocza B, Sandmaier BM. Conditioning regimens for hematopoietic cell transplantation: one size does not fit all. Blood. 2014;124(3):344–53.
24. de Sa Silva F, Almeida PN, Rettore JV, Maranduba CP, de Souza CM, de Souza GT, et al. Toward personalized cell therapies by using stem cells: seven relevant topics for safety and success in stem cell therapy. J Biomed Biotechnol. 2012;2012:758102.
25. Meyer JR. The significance of induced pluripotent stem cells for basic research and clinical therapy. J Med Ethics. 2008;34(12):849–51.
26. Liang G, Zhang Y. Genetic and epigenetic variations in iPSCs: potential causes and implications for application. Cell Stem Cell. 2013;13(2):149–59.
27. Volarevic V, Markovic BS, Gazdic M, Volarevic A, Jovicic N, Arsenijevic N, et al. Ethical and safety issues of stem cell-based therapy. Int J Med Sci. 2018;15(1):36–45.
28. Golpanian S, Schulman IH, Ebert RF, Heldman AW, DiFede DL, Yang PC, et al. Concise review: review and perspective of cell dosage and routes of administration from preclinical and clinical studies of stem cell therapy for heart disease. Stem Cells Transl Med. 2016;5(2):186–91.
29. Wong RS. Mesenchymal stem cells: angels or demons? Biomed Res Int. 2011;2011:459510.
30. Mousavinejad M, Andrews PW, Shoraki EK. Current biosafety considerations in stem cell therapy. Cell J (Yakhteh). 2016;18(2):281.
31. Goldring CE, Duffy PA, Benvenisty N, Andrews PW, Ben-David U, Eakins R, et al. Assessing the safety of stem cell therapeutics. Cell Stem Cell. 2011;8(6):618–28.

Further Reading

Bravery CA, Carmen J, Fong T, Oprea W, Hoogendoorn KH, Woda J, Burger SR, Rowley JA, Bonyhadi ML, Van't Hof W. Potency assay development for cellular therapy products: an ISCT review of the requirements and experiences in the industry. Cytotherapy. 2013;15(1):9–19.
Hayakawa T, Aoi T, Bravery C, Hoogendoorn K, Knezevic I, Koga J, Maeda D, Matsuyama A, McBlane J, Morio T, Petricciani J. Report of the international conference on regulatory endeavors towards the sound development of human cell therapy products. Biologicals. 2015;43(5):283–97.
Ito E, Miyagawa S, Takeda M, Kawamura A, Harada A, Iseoka H, Yajima S, Sougawa N, Mochizuki-Oda N, Yasuda S, Sato Y. Tumorigenicity assay essential for facilitating safety studies of hiPSC-derived cardiomyocytes for clinical application. Sci Rep. 2019;9(1):1881.

Standards and Regulatory Frameworks (for Cell- and Tissue-Based Products)

Maryam Arabi, Fereshteh Mohamadi-Jahani, Sepideh Alavi-Moghadam, and Mina Abedi

© Springer Nature Switzerland AG 2020
B. Arjmand et al. (eds.), *Biomedical Product Development: Bench to Bedside*,
Learning Materials in Biosciences, https://doi.org/10.1007/978-3-030-35626-2_9

9.1 What You Will Learn in This Chapter?

Cell therapy, as a field of biotherapeutic medicines, concerns with global public health promotion. The complexity of human cell- and tissue-based products (HCTPs) in their structure, content, mode of action, and delivery confronts health relevant professionals, regulatory authorities, and manufacturers with some challenges from product design to delivery [1].

First of all, in order to find out the difference between regulations and standards, note the following ◘ Table 9.1 [2].

This chapter, at first, introduces the importance of establishing and following standards and regulations. Then, it gives some information about the characterization of cell- and tissue-based products according to accepted standards and regulations. Finally, the following part of the chapter will provide different classifications of HCTPs, based on several aspects.

9.2 Rationale and Importance

Nowadays, biomedical discoveries, as a developing field, are being focused on. Accordingly, it introduces the translational research (the journey of biomedical discoveries from the bench to the bedside). This type of research demands high skills like individual and within-a-team ones and also an understanding of the usage of data in different regulatory frameworks [3].

In other words, the manufacturing of cell- and tissue-based products now is rapidly facing a serious challenge of assuring consistent and safe approaches in making products [1]. In biotechnological and pharmaceutical fields, basic and clinical scientists should be aware of related regulations and standards before using a cell- or tissue-based products in clinic [4]. That is why in this case talking about regulations and standards seems important.

9.3 Therapeutic Characterizations of Cell-Based Products

Because of therapeutic importance of HCTPs, some characteristics for these kinds of products can be considered:
- These products can be applicable in recipient for a lifetime.
- Their responses to their environment and other cells are not encountered with other medications.

◘ **Table 9.1** Standards vs. regulations: Although regulations and standards seem very similar, they are different in some aspects [2]

Regulations	Standards (voluntary)
Legally implemented by the governments	Frequently nongovernmental
Has the responsibility of defining safety requirements	Has the responsibility of describing how manufacturers might meet regulatory demands
Accurately informs health professionals and consumers of some specific data	Physical standards provide accepted "benchmark" materials

- According to their capacity to replicate or mature in vivo, HCTPs preserve their functionality after administration.
- They may migrate unintentionally to other tissues [1].

Because of the mentioned special characteristics of HCTPs that make them distinguishable from other pharmaceuticals, establishing sets of standards and regulations seems important. Some types of the standards and regulations are provided in the following:

9.4 Governmental Standards

Nowadays, using governmental standards is necessary because it is a growing need to consider the linkage between cost and safety, quality, and efficacy criteria in therapeutic cell-based product development. Market access of therapeutic cell-based products has to be approved by associated regulatory bodies on the basis of safety, quality, and efficacy. However, policy makers have to implement several evaluations to justify and balance cost and benefit and cost and effectiveness and generally to make factual decisions because of growing demands on public healthcare budgets. In other way, approval by the relevant regulatory authority is required for most drug reimbursement procedures either public- or insurance-based ones. In conclusion, adherence to relevant governmental standards, as a crucial prelude to significant penetration of the market, seems important [5].

9.5 Standards and Regulations in Different Countries

Regulatory agencies have developed cell- and tissue-based drug regulations. For example, the Australian framework is administered by the Therapeutic Goods Administration (TGA), the European Union by the European Medicines Agency (EMA), and the USA by the Food and Drug Administration (FDA) [3].

There are some similarities between regulations of different countries. However, some differences can be seen, basically assigned to external factors (e.g., culture) and also the expansion of their scopes within years [3].

Let us consider some of these national standards and regulatory frameworks separately.

9.5.1 The United States

The FDA Center for Biologics Evaluation and Research (CBER) and, specifically, the Office of Cellular, Tissue, and Gene Therapies (OCTGT) have the responsibility for the supervision of cellular products and tissues based on approved regulatory criteria set since the 1990s.

The point is that somatic cell therapy is considered to be "experimental" in FDA definition, rather than a standard medical practice. Therefore, cellular products cannot be used clinically without an investigational new drug (IND) application [6].

Additionally, products of tissue engineering and regenerative medicine (TE/RM) are regulated by FDA in different pathways because of the following different categories:

9.5.1.1 Tissues

FDA regulates HCTPs by rules shown in Section 361 of the Public Health Service Act (PHS Act) under the communicable disease authority that one of its major objectives is preventing contaminated and infectious tissue and cell transmission [7].

9.5.1.2 Biological Products

Tissue rules of FDA in Section 351(i) of the PHS Act name different items that can be categorized as biological products including toxin and antitoxin, virus, blood, etc. Biological products can be utilized to prevent, treat, or cure human disease or condition [8].

9.5.1.3 Medical Devices

Section 201(h) of the Federal Food, Drug, and Cosmetic Act (FD&C Act) defines medical device as recognized instrument, implement, apparatus, etc. in the official National Formulary or the US Pharmacopoeia which is considered to diagnose or treat diseases or to have effects on human body or animals, in their function or structure. Also, devices do not need metabolic pathways or chemical actions to achieve their primarily considered purposes [9].

The Medical Device Amendments of the FD&C Act categorize devices into three risk-based classes based on their indications for use. Also, it is based on their risk for patients and users. The lowest risk is for class I, and class III has the greatest risk [7].

Devices in class I are regulated by general controls. Class II requires special controls in addition to general ones, and class III needs general controls and premarket approval. Therefore, all classes require general controls. Although class III is more risky, it also requires Quality System Regulation (QS Regulation) [7, 10].

9.5.2 Canada

In January 2003, health Canada gave a document about basic needs for transplantation of human cells, tissues, and organs [11]. The Canadian regulatory framework includes guidelines, regulations, acts, and policies that all of them are involved in the regulation of cell and gene therapy products (CGTPs). In Canada, cell therapy regulation is not defined specifically and also is not included in the list of F&DR. CGTPs as pharmaceutical products can be regulated under the Food and Drugs Act [12].

9.5.3 The European Union

In the European Union, ATMPs were approved at the end of 2007 for the provision of placing viable cell-based and tissue-based products in the market which are proper for human use. Based on this legal document, both somatic cell- and tissue-based products may contain cells of human and/or animal origin, noticing this difference that in tissue-engineered products, medical devices are used in allogeneic and xenogeneic cellular treatments ex vivo or in vivo (e.g., microcapsules, intrinsic matrix scaffolds, biodegradable, or not). This document's main policy is that a centralized authorization procedure for all advanced therapy medicinal products (AMTPs) has to be set up through an interdisciplinary expert committee, the Committee for Advanced Therapies (CAT), within the European

Medicines Agency (EMEA). Based on the special safety and regulatory frameworks, in order to prevent potentially negative public health events, there will be firmer requirements on risk management, including the complete traceability of donors, recipients, tissues, and products [6].

9.5.4 Australia

Australian regulatory guidelines and also standards for biological labeling and guidelines for donor eligibility, testing, and minimizing infectious disease transmission for biologicals have been published by the Therapeutic Goods Administration (TGA). The biological regulations are categorized into different classes based on the following criteria:

- Duration of manipulating time from their extraction of naturally occurring state to its use
- The proximity of their usage to their original biological function

In conclusion, the higher degree of regulatory oversight ends to the upper class the biological will place in [13].

9.5.5 Japan

In Japan, Pharmaceuticals and Medical Devices Agency (PMDA) and the Ministry of Health, Labor, and Welfare (MHLW) are responsible for regulating cell and tissue therapy products as drugs or medical devices which have only approved autologous cultured epidermis (JACE) as a medical device until 2012 [13].

9.5.6 Korea

South Korean Food and Drug Administration (K-FDA) which is responsible for supervision of biologics or biopharmaceuticals, recombinant proteins, cell culture products, cell and gene therapy products, and others with similar materials has approved over 15 cell therapy products, namely, articular cartilage defects, cancer, burns, fracture, diabetic foot ulcer, and bone necrosis, including a stem cell product for acute myocardial infarction (Hearticellgram®-AMI) until 2012 [13].

9.5.7 Singapore

Some of the requirements in the licensing terms and conditions (LTCs) which stand for the Ministry of Health (MOH) cell- and tissue-based therapeutic (CTT) products include the following:

1. The first one is about the person's medical specialty who is in charge of the CTT facility, which has to be related to the nature of the CTT or its use.
2. Before the use of CTT products, obtaining the patients' consent is necessary.
3. Another one is about the necessity of giving a 7-day cooling-off period to the patient so that considering the CTT treatment before providing it would be possible.

4. The necessity of effective documentation system for all provided CTTs allows the authorities to maintain complete records of the CTT use, in addition to the donation, procurement, and testing of CTT products.
5. This set of criteria also describes an identification system which has the responsibility of tracking CTT products from the donor to the recipient and vice versa.
6. There are some specific conditions mentioned in this criteria set which have to be satisfied for the usage of low-evidence CTT products, which include MOH approval and attempts to all other conventional treatment options (i.e., "last resort" treatment) [13].

9.6 The Current International Standards and Regulatory Frameworks

With the purpose of discussing challenges toward the scientific regulation of HCTPs, a workshop was held in Kyoto in 2014 in which one of its recommendations was that the World Health Organization (WHO) should take an action toward beginning a work on the development of appropriate guidelines and relevant documents for National Regulatory Authorities (NRAs), National Control Laboratories, and CTP producers [1].

Moreover, in order to review the available scientific information and data and to identify and discuss the most significant current regulatory issues in the field of cell therapy, a follow-up conference was held in Tokyo in 2015 [1].

The purpose of this conference is summarized in as follows:

It is concluded that the necessary condition for the current regulatory framework to accommodate HCTPs is NRA's flexibility in the application of the regulations that they have put in place.

There were also some problems identified in this conference which include potency tests with functional readouts available that are not dependent on surrogate markers, tumorigenicity tests, and the fundamental role of international standards in controlling cell products, the point that at which and extent to which GMPs should apply.

Another conclusion was the WHO preference in order to be the organization which takes the lead in developing international guidance.

In this conference, there are also some points as follow:

At first, data of current requirements of the EU, FDA, Japan, China, and Korea are developed and were gathered to show common points and also divergence between them as a basic platform for more discussion.

In addition, based on the bone marrow transplantation (BMT) experiences, some controlling points with the aim of facilitating or inhibiting the developmental process of cell-based products should be identified.

It was also recommended to establish a rational regulatory framework for HCTPs that could be a beginning for WHO's role [1].

An international regulatory conference on human cell and gene therapy products subsequently was held in order to discuss the challenges raised at the Tokyo 2015 conference. It was held in Osaka, Japan, on March 16, 2016, and was co-sponsored by Pharmaceuticals and Medical Devices Agency (PMDA), National Institute of Biomedical Innovation, Health, and Nutrition (Japan), and the Japanese Society for Regenerative Medicine. There were also participants from regulatory agencies (e.g., FDA, Committee for Advanced Therapies/the European Medicines Agency (EMA), the European Directorate for the

Quality of Medicines and Healthcare (EDQM), Health Canada, Singapore's Health Science Authority, PMDA, Taiwan Food and Drug Authority (TFDA)), nongovernmental organizations, academia, and industry in North America, Europe, and Asia who attended the forum. In this forum, with the aim of minimizing inconsistency in regulatory approaches among interested parties and improving international regulatory convergence through an international organization such as the WHO, compiling a global best practice document was emphasized further, in order to evaluate the quality, efficacy, and safety of the CTPs [1].

There were also discussions about making a minimum consensus package (MCP) as a useful approach to the scientific evaluation of HCTPs which can be the foundation of a global guidance document. A MCP is composed of general considerations, scientific principles/concepts, and technical demands that can generally be used in the cases of most HCTPs. Regarding these points, relevant workgroups can use the MCP as a platform [1].

In order to develop a MCP, the data provided comprehensively and extensively in relevant requirements, guidelines, scientific literature, and other sources should be taken into consideration. In order to cover this need, specific, necessary scientific and technological elements for chemistry, manufacturing, and controls and nonclinical and clinical studies have to be identified and discussed further [1].

Moreover, in order to develop an individual HCTP, some other factors have to be considered such as sources of the applied cells as well as the processes of their manufacturing, product-specific profile, the procedures of product administration, specific diseases we intend to target, development stage, and experience of use, among other factors (these are called add-on packages for individual cases) [1].

In short, the forum aiming at cell therapy could be advanced efficiently, effectively, and reasonably through the utilization of such an MCP plus add-on packages for individual cases [1].

The most recent International Alliance for Biological Standardization (IABS) conference on CT was held in London on November 23, 2016. There was a discussion on key issues of manufacturing HCTPs, and there were selected aspects agreed on to inform about the draft of future guidance. As with previous IABS meetings, this one aimed at combining specific relevant aspects of industry, academia, health services, and regulatory bodies as well as making some promotions in regulation, registries, and banking of stem cell lines, requirements for raw materials, manufacturing, standardization, characterization, and preservation. Representations from 14 countries had participants in this meeting [1].

9.7 Challenges and Further Perspectives

From 1997, when the first CTPs, Carticel® (autologous cultured chondrocytes), were approved by FDA, there has been an increasing expectation that cell-based investigational products will become effective new therapies. Therefore, meeting this expectation will become a new challenge for researchers. Nevertheless, there are many areas of regulatory uncertainty and differences among countries and regions in this case that is another challenge in this field of work [1].

Other challenges are source, regulatory experience, and support limitation for running a developmental process on the road of intended products and life cycle in smaller private organizations, academic, or healthcare institutions by which many cell therapies have been developed. Therefore, in order to make it possible to effectively progress the potential new CTPs to the market, regulators need to engage early with HCTP developers [1].

Additionally, because final cell preparation for HCTP is fragile and has a limited shelf life, a specific notice should be given to the pharmaceutical pathways for autologous HCTPs that start from the donor/patient (biopsy collection) to the manufacturing site (production, quality controls, and release) [1].

Also there are some ethical considerations of xenotransplantation in cell therapy that have to be of concerned, for instance, informed consent from the recipient, the potential risks of Infections by xenotransplantation, the social acceptance of this kind of transplantation, as a solution to human organ shortage, and the use of animals, especially primates as recipients in preclinical trials of xenotransplantation [6].

Moreover, there are some challenges in all transplantation, i.e., consideration of focusing on specific procedures especially informed consent, patient quality of life, benefits versus risks, the recipient autonomy, and medical paternalism or nonmaleficence [14].

Take a look at the following example of xenotransplantation challenges in the UK:

Box 9.1

There are some regulatory concerns in using primates in the UK focusing on the uncertainty about the acceptability of this procedure in the public. That is because of their evolutionary proximity to human beings [6].

As the Weatherall Report, published at the end of 2006, emphasized, the UK expert group accepted nonhuman primate research with specific conditions, including high quality research with the potential to benefit mankind, and if it is the only way of solving important scientific or medical questions [6].

Take-Home Messages

- Regulations provide governmental status about safety requirements and accurate information for health professionals.
- Standards are often formed outside of government and talk about how manufacturers might cover regulatory demands.
- Basic and clinical scientists should be aware of related regulations and standards in their field as manufacturing of cell- and tissue-based products is rapidly developing.
- Regulations in tissue engineering and regenerative medicine (TE/RM) are categorized into different pathways including tissue, biological products, and medical devices.
- Several countries like the USA, Canada, Australia, Singapore, Japan, and Korea have designed their own regulatory policies.

References

1. Petricciani J, Hayakawa T, Stacey G, Trouvin J-H, Knezevic I. Scientific considerations for the regulatory evaluation of cell therapy products. Biologicals. 2017;50:20–6.
2. George B. Regulations and guidelines governing stem cell based products: clinical considerations. Perspect Clin Res. 2011;2(3):94–9.

3. Ilic N, Savic S, Siegel E, Atkinson K, Tasic L. Examination of the regulatory frameworks applicable to biologic drugs (including stem cells and their progeny) in Europe, the U.S., and Australia: part I-A method of manual documentary analysis. Stem Cells Transl Med. 2012;1(12):898–908.
4. Halme DG, Kessler DA. FDA regulation of stem-cell-based therapies. N Engl J Med. 2006;355(16): 1730–5.
5. Farrugia A. When do tissues and cells become products? – regulatory oversight of emerging biological therapies. Cell Tissue Bank. 2006;7(4):325–35.
6. Tallacchini M, Beloucif S. Regulatory issues in xenotransplantation: recent developments. Curr Opin Organ Transplant. 2009;14(2):180–5.
7. Lee MH, Arcidiacono JA, Bilek AM, Wille JJ, Hamill CA, Wonnacott KM, et al. Considerations for tissue-engineered and regenerative medicine product development prior to clinical trials in the United States. Tissue Eng Part B Rev. 2010;16(1):41–54.
8. Spring S. Regulatory considerations for human cells, tissues, and cellular and tissue-based products: minimal manipulation and homologous use. Available from: https://www.fda.gov/regulatory-information/search-fda-guidance-documents/regulatory-considerations-human-cells-tissues-and-cellular-and-tissue-based-products-minimal.
9. Classification of products as drugs and devices and additional product classification issues: Guidance for industry and FDA staff. US Food and Drug Administration. 2017. http://www.fda.gov/CombinationProducts/default.htm. Accessed May 4, 2019.
10. Classify your medical device. US Food and Drug Administration. 2014. https://www.fda.gov/medical-devices/overview-device-regulation/classify-your-medical-device. Accessed May 4, 2019.
11. Lloyd-Evans M. Regulating tissue engineering. Mater Today. 2004;7(5):48–55.
12. Chisholm J, Ruff C, Viswanathan S. Current state of Health Canada regulation for cellular and gene therapy products: potential cures on the horizon. Cytotherapy. 2019;21(7):686–98.
13. Kellathur SN, Lou H-X. Cell and tissue therapy regulation: worldwide status and harmonization. Biologicals. 2012;40(3):222–4.
14. Simmons PD. Ethical considerations in composite tissue allotransplantation. Microsurgery. 2000;20(8):458–65.

Further Reading

Articles

Hayakawa T, Harris I, Joung J, Kanai N, Kawamata S, Kellathur S, et al. Report of the international regulatory forum on human cell therapy and gene therapy products. Biologicals. 2016;44(5):467–79.

Online Sources

https://www.iabs.org/.

Principles of Good Clinical Practice

Najmeh Foroughi-Heravani, Mahdieh Hadavandkhani, Babak Arjmand (ID)*, Bagher Larijani, Alireza Baradaran-Rafii, Ensieh Nasli-Esfahani, and Moloud Payab*

The original version of this chapter was revised. The correction to this chapter can be found at
https://doi.org/10.1007/978-3-030-35626-2_14

10.1 What You Will Learn in This Chapter

The design, monitoring, recording, analyzing, and reporting of clinical trials are major issues which need to be standardized all over the world. Accordingly, good clinical practice (GCP) is an international ethical and scientific quality standard for conducting the trials involving human subjects. In this chapter, a brief history of GCP, leading to its importance and functions, will be discussed. Then, we will go through some details about its principles and some major roles of components including the patients, sponsor, investigator, and other stakeholders. Before addressing the challenges which you will be faced with using the guidelines, there are some examples about different kinds of GCP guidelines around the world to show how it is implemented and influences the performance and management of clinical trials.

10.2 What Is Good Clinical Practice?

Good clinical practice (GCP) is defined as a "standard for the design, conduct, performance, monitoring, auditing, recording, analyzing and reporting of clinical trials." It provides assurance that the obtained data and reported results are credible and accurate. On the other hand, the rights, integrity, and confidentially of trial subjects are protected by GCP [1].

10.3 History of GCP

Prior to actual set of guidelines to follow for GCP, clinical studies could result in serious diseases and even death which made them quite dangerous. In 1906, some harmful drugs that could be bought over the counter led to the first landmark in the regulation of drugs as Pure Food and Drugs Act [2]. Then, in 1930, the Food and Drug Administration (FDA) (first known as the Food, Drug, and Insecticide Administration in 1927) was established to oversee compliance with the Pure Food and Drugs Act. However, it was not much successful in preventing all unsafe drugs from being applicable, and when the Federal Food, Drug, and Cosmetic Act was released by the US Congress in 1938, it was for the first time that the manufacturers were demanded to test the safety of drugs and demonstrate the evidence to the FDA before marketing [2].

Following the unethical experiments performed in Germany during World War II (WWII), the Nuremberg Code was developed in 1947 which expressed the need for a scientific basis in research on humans. This code arranged ethical guidelines and stated the necessity of informed consent [2, 3].

In response to the severe fetal limb deformities related to the use of maternal thalidomide, the Kefauver-Harris Amendments were passed in 1962 which made the evaluation of all new drugs for safety and efficacy a responsibility for the FDA. Another turning point occurred in 1964 by the World Medical Association (WMA) that introduced Declaration of Helsinki which formed the basis for the ethical principles underlying the International Conference on Harmonization (ICH)-GCP guideline. It was a set of recommendations for guiding medical doctors in biomedical research including human subjects. Although it influenced national legislations, there was no international standard for harmonization and globalization [1].

The Belmont Report issued in 1979 by the National Commission for Protection of Human Subjects of Biomedical and Behavioral Research is another milestone in the formation of ICH-GCP guidelines [1]. This report identified three fundamental principles: (1) Respect for persons: Obtaining informed consent; (2) Beneficence: Assessment of risks and potential benefits; and (3) Justice: Equitable selection and fair treatment.

In this context, the World Health Organization (WHO) and the Council for International Organization of Medical Sciences (CIOMS) issued the "Proposal International Guidelines for Biomedical Research Involving Human subjects" in 1982 to make it easier for developing countries to apply the Declaration of Helsinki and the Nuremberg Code. It was again revised and resulted into the "International Guidelines for Biomedical Research Involving Human Subjects" [2]. Eventually, the ICH issued the "ICH Guidelines: Topic E6 Guideline for GCP" in May 1996 to overcome international GCP variability throughout countries [2]. The guidelines, which were implemented from 1997, present a comprehensive and global guidance of the design and conduct of clinical trials. A summary of the mentioned milestones during the development of GCP is shown in ◘ Fig. 10.1 [1–3].

10.4 Rationale and Importance

Looking at the process of events which led to formation of GCPs reveals the importance of these guidelines. As clinical trials are the gold standard for proving the safety and effectiveness of new therapies, they need to be conducted based on GCP guidelines [4]. Accordingly, GCP ensures the accuracy, verifiability, and reproducibility of the generated data, enhances the ethical awareness in the clinical studies, secures the privacy of human subjects, and increases competition while improving trial methods. It helps comprehend the clinical trial significance better, clarify double-sided recognition of information, and bring political and public concern over safety aspects. Further, GCP is necessary for managing the accidents and adverse events which occur during the study [1]. Moreover, the enactment of GCP has caused reduction of costs for pharmaceutical companies during developmental processes [3]. All in all, GCP can be defined as ethics plus quality data [5].

10.5 Principles of GCP Guidance

GCP guidelines are mainly focused on the protection of human rights and providing assurance that the newly developed products are safe. Also, they present standards on how clinical trials should be conducted. Guidelines define the roles and responsibilities of clinical sponsors, clinical research associates, and monitors [4, 6].

The 13 core principles of ICH-GCP are as follows:
1. Clinical trials should be conducted in accordance with the Declaration of Helsinki, GCP, and regulatory requirement(s).
2. Before starting the trial, potential risks should be weighed against foreseeable benefits for the subjects and society. Only if the benefits justify the risks the trial is allowed to be commenced.
3. The most essential considerations which should have priority over interests of science and society are the rights, safety, and well-being of the human subjects.

10

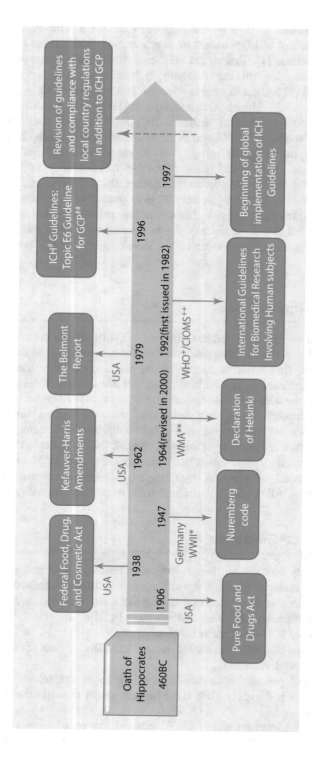

Fig. 10.1 Milestones in the development of GCP. The concept of the "good physician" is not a new matter of concern; it dates back to the ancient world as it is evidenced by the Hippocratic Oath (460 BC) [1]. The lethal and harmful drugs lead to serious official actions first known in the USA and then around the world to protect the human rights and safety. The developmental process leading to creation of ICH-GCP guidelines and some of the remarkable turning points during this evolutional road are exhibited above. Also, it can be seen that the journey is not finished yet and new revisions and local guidelines are made and utilized since these efforts were made. *World War II, **World Medical Association, +World Health Organization, ++Council for International Organization of Medical Sciences, #International Conference on Harmonization, ##Good Clinical Practice

4. There should be sufficient clinical and nonclinical information on an investigational product available to support the proposed clinical trial.
5. Scientific clinical trials should be described clearly and comprehensively in a protocol.
6. The protocol, which the trial is conducted in accordance with, must first receive institutional review board (IRB)/independent ethics committee (IEC) approval/favorable opinion.
7. A qualified physician takes the responsibility for the medical care or the decisions related to the subjects.
8. The individuals involved in a trial should be totally educated, trained, and experienced to do their responsibilities correctly.
9. Before participating in a trial, voluntary informed consents should be obtained from the participants.
10. The information of the trial should be recorded and managed to make its reporting and authentication available.
11. Protection of the privacy and confidentiality of records identifying the subjects is of great importance which should be in consistent with the relevant rules of the applicable regulatory requirement(s).
12. Good manufacturing practice (GMP) leads the way in producing, handling, and storing the investigational products. Also, the products should be used in accordance with the approved protocol.
13. Special systems with procedures that guarantee the details of the trial should be implemented.

10.6 Major Stakeholders of the GCP Protocol

Management, implementation, and taking charge of GCP guidelines lie with the stakeholders. Hence, their responsibilities should be introduced as follows:

- Regulatory authorities: The authorities that review submitted clinical data and those that administer inspections.
- The sponsor: An individual (company, institution, or organization) which is responsible for the initiation, management, and financing of a clinical trial.
- The project monitor: Acts as the main line of communication between the sponsor and the investigator and is usually appointed by the sponsor.
- The investigator: A team leader who is responsible for the conduct of the trial at the trial site.
- Subject/trial subject: An individual who participates in a clinical trial, either as a recipient of the investigational product(s) or as a control.
- The pharmacist at the trial location: Responsible for maintenance, storage, and supplying of investigational products such as drugs in clinical trials.
- Ethical review board or committee for protection of subjects: Assigned by institution or "if not available then the Authoritative Health body in that country will be responsible."
- Committee to monitor large trials: Oversees sponsors such as pharmaceutical companies [1, 6].

The definition and responsibilities of main ones are discussed in ► Chap. 12.

10.7 GCP around the World

The ICH-GCP was designed to promote harmonization in the three prominent ICH regions, the USA, Japan, and Europe. Also, the WHO-GCP was established for less accomplished pharmaceutical agencies in nations with no other guidelines [3].

The Office of Good Clinical Practice (OGCP) is the focal point within the FDA for GCP and Human Subject Protection (HSP) issues that may arise in clinical trials regulated by FDA [7]. FDA regulations govern the conduct of clinical trials and describe GCP for studies (human and non-human animal subjects). Through different Codes of Federal Regulations (CFR), some essential issues are discussed such as electronic records, electronic signature (21 CFR part11), regulatory hearing before the food and drug administration (21 CFR part16), protection of human subjects (informed consent) (21 CFR part50), and applications for FDA approval to market a new drug (21 CFR part314) [5, 7].

In Europe, Clinical Trial Directive (Directive 2001/20/EC) was introduced to simplify and coordinate the administrative provisions governing clinical trials [2, 8].

Clinical research is an example of the huge disparities between high-income countries (HICs) and low- and middle-income countries (LMICs) in access to adequate care [9]. After the conception of ICH-GCP guidelines, various countries in the Asia-Pacific region formulated their own guidelines such as Singapore GCP, Chinese GCP, Malaysian GCP, and also Iranian GCP. The Malaysian Guideline for GCP was first published in 1999 which adopts the basic principles outlined by the ICH-GCP with some modifications to suit local requirements. In India, the first recognized document for GCP was formulated in 2001 by the Central Drugs Standard Control Organization (CDSCO). The composition of an Ethics Committee is specified in the Indian GCP [1, 3]. All in all, GCP is now an international standard, based on which enormous clinical trials with the cooperation of multiple countries and enrollment of participants from different nationalities can be conducted.

10.8 Challenges and Future Direction

GCP principles, commonly applied to clinical researches on human subjects, serve "as a roadmap of responsibilities" for the individuals involved in research [10]. Conducting clinical trials based on GCP causes many problems. For instance, there is the lack of commitment of clinicians due to having little time for clinical research as they have lots of other responsibilities [11]. Even, there are a limited number of GCP-trained competent professionals which pose a major challenge in the implementation of GCP guidelines. Professional training on GCP across various stakeholders (sponsor, investigator, patient, regulators, etc.) is a crucial requirement. Accordingly, FDA conducts GCP training which recently is available online [12]. Not only is there a lack of supportive infrastructure like labs and diagnostics, but also most hospitals do not meet the infrastructure requirements as per GCP guidelines. The upgrade of local laboratories and the harmonization of laboratory quality systems are known as major challenges in LMICs [9].

Although the intention of GCP is to protect and promote patients' rights and safety and increase the overall quality of clinical trials, it also has some deficiencies or somehow unnecessary issues which result in challenge themselves. For example, some GCP processes can lead to markedly increased trial complexity, duration, and costs while they do not significantly improve the quality of these trials and their ability to correctly answer clinical questions or do not support the safety of human subjects. In some circumstances, GCP guidelines put inappropriate emphasis on reporting like progress reports and safety reporting. The guidance is derived from informal concurrence rather than evidence-based data. Implementation of GCP guidelines requires intensive site monitoring which is known to be costly and time consuming. The main problem is that there is an emphasis on documents unrelated to research validity (e.g., updating curriculum vitae) rather than the specifically designed activities for improvement of research quality based on the major goal of monitoring, the detection of inaccurate transcription of data [10].

Concerning laboratory quality management systems, GCP codes do not pay much attention for the laboratories of the trial, and they do not refer to the Good Clinical Laboratory Practices (GCLP) code. GCP guidelines should provide a section describing the qualification and responsibilities and monitoring the trial laboratories based on GCLP. The quality of investigational medicinal product is another essential part which requires compliance of the international GCP codes with locally applicable GMP. Since degraded, underdosed, and nonbioequivalent medicines have been used in trials, more detailed evidences of the inspecting and reporting of the quality of drugs and medical devices are needed [9].

Along with the increase amount of utilization of GCP guidelines around the world, more and more clinical trials will be designed and conducted in the future which are expected to comply with GCP codes. Besides, it is suggested that there is the need to revise the guidelines to improve relevance to the current research environment as these guidelines are "meant to serve the interests of global society" [9, 10].

Take-Home Messages

- GCP is an international ethical and scientific quality standard for conducting the trials.
- GCP provides assurance that the data and reported results are credible and accurate and that the rights and integrity of trial subjects are protected.
- GCP has 13 core principles which mainly focus on human rights and verifiability of the data. Other key aspects are justifying the risks, regarding the regulatory requirements and the protocol, providing adequate recording and reporting of the data, involving qualified individuals, etc.
- The ICH-GCP is not the only guideline available, and there are other national and international GCP codes such as Clinical Trial Directive of Europe, Malaysian guideline, and Indian guideline.
- The implementation of GCP poses obstacles and challenges like the lack of trained staff and required infrastructure, which arise worries.
- Although there are international guidelines such as ICH-GCP and WHO-GCP which revolutionized the process of clinical research, there are still serious challenges. Therefore, some revisions and improvement are needed.

References

1. Vijayananthan A, Nawawi O. The importance of good clinical practice guidelines and its role in clinical trials. Biomed Imaging Intervention J. 2008;4(1):e5.
2. Otte A, Maier-Lenz H, Dierckx RA. Good clinical practice: historical background and key aspects. Nucl Med Commun. 2005;26(7):563–74.
3. Narayanan VA, Fernandes S, Sandeep DS, Kumar P, Castelino L. Good clinical practices: an Indian perspective. Res J Pharm Technol. 2018;11:3209.
4. American Society of Clinical Oncology. Good clinical practice research guidelines reviewed, emphasis given to responsibilities of investigators: second article in a series. J Oncol Pract. 2008;4(5):233–5.
5. "Jao" Lacorte L. Good clinical practice 101: an introduction. CDRH learn. 2010. https://www.slideshare.net/gy2139/cdrh-learn-module-gcp-101-lacorte. Accessed 12 May 2010.
6. Integrated Addendum to ICH E6(R1): Guideline for Good Clinical Practice E6(R2). 2016. https://www.ich.org/products/guidelines/efficacy/efficacy-single/article/integrated-addendum-good-clinical-practice.html . Accessed 1 Mar 2018.
7. U.S. Food and Drug Administration: Office of Good Clinical Practice. https://www.fda.gov/about-fda/office-special-medical-programs/office-good-clinical-practice. Accessed 19 June 2019.
8. European Medicins Agency. https://www.ema.europa.eu/en.
9. Ravinetto R, Tinto H, Diro E, Okebe J, Mahendradhata Y, Rijal S, et al. It is time to revise the international Good Clinical Practices guidelines: recommendations from non-commercial North–South collaborative trials. BMJ Glob Health. 2016;1(3):e000122.
10. Mentz RJ, Hernandez AF, Berdan LG, Rorick T, O'Brien EC, Ibarra JC, et al. Good clinical practice guidance and pragmatic clinical trials: balancing the best of both worlds. Circulation. 2016;133(9):872–80.
11. Merican MI. Good clinical practice: issues and challenges. Med J Malaysia. 2000;55(2):159–63.
12. U.S. Food and Drug Administration. https://www.fda.gov.

Further Reading

Books

Allport-Settle MJ. Good clinical practice: pharmaceutical, biologics, and medical device regulations and guidance documents concise reference; Volume 1, Regulations. 1st ed. Willow Springs: PharmaLogika; 2010.
Hutchinson D. Essential good clinical practice. 1st ed. Guildford: Canary Ltd; 2009.
McGraw MJ, George AN, Shearn SP, Thomas F, Haws TF Jr, Hall RL. Principles of good clinical practice. 1st ed. London/Chicago: Pharmaceutical Press; 2010.

Online Resources

Integrated Addendum to ICH E6(R1): Guideline for Good Clinical Practice E6(R2). 2016: https://www.ich.org/products/guidelines/efficacy/efficacy-single/article/integrated-addendum-good-clinical-practice.html.
U.S. Food and Drug Administration: https://www.fda.gov.
WHO-GCP guideline: https://apps.who.int/medicinedocs/en/d/Jwhozip13e/7.4.html.

10

Design, Performance, and Monitoring of Clinical Trials

Mahdieh Hadavandkhani, Najmeh Foroughi-Heravani, Parisa Goodarzi, Moloud Payab, Hamid Reza Aghayan, Sepideh Alavi-Moghadam, Mehrnoosh Yarahmadi, Motahareh Sheikh-Hosseini, and Bagher Larijani

The original version of this chapter was revised. The correction to this chapter can be found at https://doi.org/10.1007/978-3-030-35626-2_14

© Springer Nature Switzerland AG 2020
B. Arjmand et al. (eds.), *Biomedical Product Development: Bench to Bedside*,
Learning Materials in Biosciences, https://doi.org/10.1007/978-3-030-35626-2_11

11.1 What You Will Learn in This Chapter?

You will first understand the concept and definition of the clinical trials. Then, the importance and necessity of its implementation will be described. At the next step, this chapter tries to clarify the importance of phases of clinical trials (phases III and IV). In this part, you should pay attention to the type of biomedical studies including pharmaceutical, cell, gene, and tissue investigations which are mentioned. You will learn about fundamentals of clinical trials after their clarification. According to international standards and guidelines, inclusion and exclusion criteria of eligible participants will be demonstrated, and intervention includes concepts like diagnostic intervention. Furthermore, a snapshot of clinical trial monitoring, risk assessment, and its various types will be exhibited with a brief focus on registration and obstacles.

11.2 What Are Clinical Trials?

Investigational new drugs (INDs) can be available to the public after a hardworking stepwise study and proof of safety and efficacy. As a result, INDs should be evaluated in a scientific investigational setting. Accordingly, clinical trial is one of the important gold standard tests. It has a protocol for performing trials which illustrate several aspects of studies such as sex, age, etc. In the end, INDs (novel drugs and treatments) will be available to patients after passing safety and efficacy stages. Logically, all of these stages are under supervision of international, national, and regional authorized bodies (e.g., FDA, KFDA,). In summary, the purpose of conducting clinical trials is to improve medical and behavioral intervention [1, 2].

11.3 Rationale and Importance

Clinical trials play an important role in discovering novel treatments. They help scientists to identify and diagnose various diseases. On the other hand, clinical trials can provide valuable infrastructure for treatment of several diseases and decreasing the complications. In other words, factual and documented results which are extracted from clinical trials are valuable in patient care. Finally, using clinical trials, investigators and doctors can assess the risks of new drugs against their benefits and then decide to whether prescribe it or not [3].

11.4 Phases of Clinical Trial

In a clinical point of view, a novel treatment needs to pass four phases of clinical trials (◻ Fig. 11.1).

11.4.1 Phase I

The main aim of this phase is to illustrate the best dose and method with an acceptable level of risk (safety phase). In this phase, researchers evaluate probable risks and side

Fig. 11.1 Phases of clinical trial. Phase I: Takes weeks, aim to IND human safety, tens of participants (up to 50). Phase II: Takes months, efficacy examination, hundreds of participants (up to 500). Phase III: Takes several years, result confirmation, thousands of participants. Phase IV: Long term, health authority surveillance and observation, massive population

effects of the INDs. Enrolling participants in this phase depends on the characteristics of the studies, and the maximum number is up to 50 (e.g., in some cases such as cell-based and gene therapies, the total number of participants is around 10 people). Physicians use a few dose for a few patients at the early step of phase I. After that, the dose of INDs increases along with the number of subjects gradually. This process continues until the side effects are very intense or the desired effect is seen. Finally, phase I steps at this stage. This drug may help patients (efficacy), but the purpose of this phase is to check the human safety of the drug. If the drug or approach is sufficiently human safe, it can be tested in a clinical trial of phase II [4].

11.4.2 Phase II

In phase II, the main focus is on evaluating efficacy. There are significant differences between this phase and phase I which include the following: (1) phase II is longer and it takes months. (2) The number of participants of phase II is more than phase I. Of course, the number of patients depends on the type of biomedical study (e.g., pharmaceutical or cell-based study). At this stage, the new developed products are compared with standard of care ones. Then, it will enter phase III if the efficacy and safety are acceptable [5].

11.4.3 Phase III

In phase III, the side effects and efficacy of INDs are assessed. In this phase, lots of drugs and methods which were considered successful do not pass due to the detrimental side effects and lack of therapeutic effects. In phase III, clinical trial investigators compare INDs with the standard of care to determine which ones work better [6].

11.4.3.1 Basic Principles

This phase has several conditions and includes the following:

1. Participants in study are alike, and there is no difference between them. They should be adjusted based on age, gender, etc.
2. The research is randomized and patients are chosen incidentally.
3. The level of efficacy is associated with either single- or double-blind process.
4. Every patient should be observed precisely [7].

11.4.3.2 Features of the Participants in Trial Groups

Phase III has 2 significant features:

- The large amount of participants (up to thousands)
- Control group such as scheme group

For election of scheme (control group), some components including objective, amount of participants, and steps multiplicity play important roles. The essential purpose of phase III is providing new and better approach for prevention, management, and treatment of diseases. The number of patients included in phase III varies based on the type of study. In pharmaceutical studies, more than 1000 patients can be enrolled, whereas in cell-, gene-, and tissue-based studies, this number significantly drops [8].

11.4.3.3 Premature Termination

The study terminates before the deadline under two conditions:

1. If the adverse events of them in one or more groups are too intense
2. If one or more groups culminate in much better results [9].

11.4.4 Phase IV

After phase III, new drugs and treatments enter the market. If they have unknown adverse effects, they will be excluded. Phase IV trials evaluate the safety of INDs in typical patients approved by the health authorities. The drug is tested in large amount of patients in long-term surveillance, which allows finding diverse effects in large groups of people. Moreover, it is helpful for doctors to know more information about the IND's works [3].

11.5 Statistics in Clinical Trial Design

Clinical trial design needs careful planning with statistical methods including the following:

11.5.1 Alternative Hypothesis

Alternative hypothesis claims that researchers would like to make it at the end of the trials. It is stated to evidence something (e.g., the IND is more effective than the conventional treatments), so a complement (called the null hypothesis) is assumed true and then they seek contraindicatory evidence. If the contraindicatory evidence is observed, the desired claim (the alternative hypothesis) will be proved, and the compliment will be disproved [10].

11.5.2 Null Hypothesis

This claim is enounced when there is no connection between groups. Null hypothesis and alternative hypothesis are opposites (pay attention to "alternative hypothesis") [10, 11].

11.5.3 Intent to Treat (ITT)

It is useful for running a randomized clinical trial (RCT) and analyzing data. The plan implies careless of adherence or treatment received. It can afford unbiased comparisons among the trial groups [10, 12].

11.5.4 Type I Error (α)

It is known as the likelihood of incorrect rejection of the true null hypothesis or false-positive conclusion. For example, INDs should cure a disease, but substantially it does not [10].

11.5.5 Type II Error (β)

It is defined as the acceptance of false null hypothesis or false-negative finding. For example, in ascendancy trials, it means the likelihood of failing to recognize a treatment effect when indeed a true treatment effect exists [10].

11.6 Eligibility Criteria and Intervention

11.6.1 Inclusion and Exclusion Criteria

The inclusion and exclusion criteria can direct the enrolment of participants based on relevant standards. These criteria contain several items such as gender, age, race, ethical principles, type of disease, medical history, mental status, and other medical situations. They play an important role in assuring patient safety during trials, justification of subjects, decrease cost, and side effects of new treatments [2].

11.6.2 Intervention

The intervention in clinical trials aims to evaluate the safety and efficacy of INDs. Accordingly, there are several categories of interventions including the following [3, 13]:

11.6.2.1 **Treatment**
Examination of INDs for various therapeutic purposes

11.6.2.2 **Prevention**
Investigation of new methods for improving lapse or relapse of diseases (e.g., vaccines, lifestyle change, etc.)

11.6.2.3 **Diagnostics**
Discovering higher methods for diagnosing type and stage of diseases

11.6.2.4 **Supportive Care**
Exploration of high quality and comfortable lifestyle for subjects with a long-term chronic condition

11.6.2.5 **Health Services Research**
Enhancing the delivery, process, management, organization, or financing of healthcare system

11.6.2.6 **Basic Science**
Focusing on the function of interventions

11.7 **Clinical Trial Monitoring**

Although there are several standards and regulations for harmonizing the framework of clinical trials, the performance of the experiments may be different according to the specific circumstances of each study. Monitoring objectives focus on the safety and welfare of the participants and also the protection of patients' rights at all stages of testing, collecting, and analyzing data. Furthermore, monitoring has several types such as on-site, oversight, and off-site methods [14].

11.7.1 **Risk Assessment**

Risk assessment is made of the following several steps [15]:

11.7.1.1 **Risk Identification**
The first step for beginning the risk assessment is answer to the questions educed from ICH Q9 including the following:
— "What might go wrong?"
— "What is the probability it will go wrong?"
— "What are the consequences (severity)?"

11.7.1.2 **Risk Analysis**
It includes examination of the likelihoods, causes, and results of risk occurrence.

11.7.1.3 Risk Evaluation

For evaluation, the risks of clinical trial researchers should answer ICH Q9 questions as follows:

- "What is the acceptable level of risk for the clinical study?"
- "Is the risk above an acceptable level?"
- "What can be done to reduce or eliminate risks?"
- "Are new risks introduced as a result of the identified risks being controlled?" [16, 17].

11.7.2 Clinical Trials Registration

The first clinical trials registration was started in 1994 in the WHO's International Clinical Trials Registry Platform (ICTRP) databases. After that, it was improved between 2004 and 2015 noticeably. Registration is essential for clearness of information and enhancing the scientific and ethical profits. Nowadays, several aspects of clinical trials can be found in registration databases (e.g., "primary registry and trial identifying number, date of registration in primary registry, secondary identifying numbers, source(s) of monetary or material support, primary sponsor, secondary sponsor(s), contact for public questions, contact for scientific queries, public title, scientific title, countries of recruitment, health condition(s) or problem(s) studied, intervention(s), key inclusion and exclusion criteria, study type, date of first enrolment, recruitment status, primary outcome(s), and key secondary outcomes" are the most important data based on WHO Trial Registration Data Set (TRDS)). Additionally, target sample size, status of clinical trials, and conditions of enrolment are other features that should be registered by investigators [18, 19].

11.8 Challenges and Future Direction

11.8.1 Eligible Patient Enrollment

One of the significant barriers is about enrolling the patients. The process of choosing patients with the acceptable conditions to engage in the researches is very difficult, and institutions or universities play an important role in facilitating this process. Afterwards, more barriers will arise. Since some of appropriate participants are reluctant to take part in the study, they will be excluded during the trial [20].

11.8.2 During the Trials

11.8.2.1 Cost of Trials

The provision of welfare and medical facilities for patients, the cost of providing tools, equipment, and staff salaries may be more than the budget for the study purposes [21].

11.8.2.2 Long Time Span

Accordingly, long time span (an average of 5–7 years) poses the studies to financial and safety management. During the trials, participants must be observed; hence, long-term observation needs a lot of staff and equipment [22–24].

Take-Home Messages

- Clinical trials are the gold standard tests for evaluating INDs.
- The goals of phases I and II are examination of human safety and efficacy.
- In phase III clinical trials, investigators compare INDs with the standard of care.
- Phase IV trials evaluate the safety of INDs in typical patients approved by the health authorities.
- Careful planning with statistical methods helps the researchers to decrease the errors.
- The types of intervention include treatment, prevention, diagnostics, supportive care, health service research, and basic science.
- Clinical trial should be done in a standard framework so it needs monitoring, risk assessment, and registration.
- Clinical trials like other kinds of researches have some barriers and challenges. Note that if the side effects are irreparable, it must be terminated.

References

1. Zelen M. A new design for randomized clinical trials. N Engl J Med. 1979;300(22):1242–5.
2. Prentice RL. Surrogate endpoints in clinical trials: definition and operational criteria. Stat Med. 1989;8(4):431–40.
3. ClinicalTrials.gov. https://clinicaltrials.gov/.
4. O'Quigley J, Pepe M, Fisher L. Continual reassessment method: a practical design for phase 1 clinical trials in cancer. Biometrics. 1990;46(1):33–48.
5. Shan G, Wilding GE, Hutson AD, Gerstenberger S. Optimal adaptive two-stage designs for early phase II clinical trials. Stat Med. 2016;35(8):1257–66.
6. Buckstein R, Meyer RM, Seymour L, Biagi J, MacKay H, Laurie S, et al. Phase II testing of sunitinib: the National Cancer Institute of Canada clinical trials group IND program trials IND. 182–185. Curr Oncol. 2007;14(4):154.
7. Nottage M, Siu LL. Principles of clinical trial design. J Clin Oncol. 2002;20(18; SUPP):42s–6s.
8. White CL, Szychowski JM, Roldan A, Benavente M-F, Pretell EJ, Del Brutto OH, et al. Clinical features and racial/ethnic differences among the 3020 participants in the Secondary Prevention of Small Sub-cortical Strokes (SPS3) trial. J Stroke Cerebrovasc Dis. 2013;22(6):764–74.
9. Bourin M. Phase III clinical trials: what methodology. SOJ Pharm Pharm Sci. 2017;4(4):1–3. http://dx.doi.org/10.15226/2374-66/4/4/00165.
10. Evans SR. Fundamentals of clinical trial design. J Exp Stroke Transl Med. 2010;3(1): 19–27.
11. Blackwelder WC. "Proving the null hypothesis" in clinical trials. Control Clin Trials. 1982;3(4):345–53.
12. A 14-month randomized clinical trial of treatment strategies for attention-deficit/hyperactivity disorder. The MTA Cooperative Group. Multimodal Treatment Study of Children with ADHD. Arch Gen Psychiatry. 1999;56(12):1073–86.
13. Ross J, Tu S, Carini S, Sim I. Analysis of eligibility criteria complexity in clinical trials. Summit Transl Bioinform. 2010;2010:46.
14. Molloy SF, Henley P. Monitoring clinical trials: a practical guide. Trop Med Int Health (TM & IH). 2016;21(12):1602–11.

15. Higgins JP, Altman DG, Gøtzsche PC, Jüni P, Moher D, Oxman AD, et al. The Cochrane Collaboration's tool for assessing risk of bias in randomised trials. BMJ. 2011;343:d5928.
16. Guideline on risk management for clinical research. European clinical research infrastructures network integrating activity. Version 1.0. 2015. https://ecrin.org. Accessed March 12, 2019.
17. ICH guideline Q9 on quality risk management. European Medicines Agency. 2015;1–20.
18. Viergever RF, Li K. Trends in global clinical trial registration: an analysis of numbers of registered clinical trials in different parts of the world from 2004 to 2013. BMJ Open. 2015;5(9):e008932.
19. Viergever RF, Ghersi D. The quality of registration of clinical trials. PLoS One. 2011;6(2):e14701.
20. Ha C, Ullman TA, Siegel CA, Kornbluth A. Patients enrolled in randomized controlled trials do not represent the inflammatory bowel disease patient population. Clin Gastroenterol Hepatol. 2012;10(9):1002–7.
21. Patsopoulos NA. A pragmatic view on pragmatic trials. Dialogues Clin Neurosci. 2011;13(2):217.
22. Winkfield KM, Phillips JK, Joffe S, Halpern MT, Wollins DS, Moy B. Addressing financial barriers to patient participation in clinical trials: ASCO policy statement. J Clin Oncol. 2018;Jco1801132.
23. Martin PS. Barriers to patient enrollment in therapeutic clinical trials for cancer: a landscape report. 2018. https://www.acscan.org/policy-resources/clinical-trial-barriers. Accessed March 20, 2019.
24. Clark LT, Watkins L, Pina IL, Elmer M, Akinboboye O, Gorham M, et al. Increasing diversity in clinical trials: overcoming critical barriers. Curr Probl Cardiol. 2019;44(5):148–72.

Further Reading

Online Resources

FDA: https://www.fda.gov
The ICH-GCP guideline: https://www.ema.europa.eu/en/ich-e6-r2-good-clinical-practice

Books

Brody T. Clinical trials: study design, endpoints and biomarkers, drug safety, and FDA and ICH guidelines. 2nd ed. Amsterdam: Academic Press; 2016.

Fayers P, Hays R, editors. Assessing quality of life in clinical trials: methods and practice. 2nd ed. Oxford: Oxford University Press; 2005.

Machin D, Day S, Green S, editors. Textbook of clinical trials. Chichester: Wiley; 2007.

Meinert CL. An insider's guide to clinical trials. New York/Oxford: Oxford University Press; 2011.

Meinert CL. Clinical trials: design, conduct, and analysis. 2nd ed. New York/Oxford: Oxford University Press; 2012.

O'Kelly M, Ratitch B. Clinical trials with missing data. Chichester: Wiley; 2014.

Rosenberger WF, Lachin JM. Randomization in clinical trials: theory and practice. 2nd ed. Hoboken: Wiley; 2016.

Smith PG, Morrow RH, Ross DA. Field trials of health interventions: a toolbox. 3rd version. Oxford: Oxford University Press; 2015.

Wang D, Bakhai A. Clinical trials: a practical guide to design, analysis, and reporting. 1st ed. London: Remedica; 2007.

11

Good Clinical Practice: Guidelines and Requirements

Najmeh Foroughi-Heravani, Mahdieh Hadavandkhani, Babak Arjmand (ID)*, Moloud Payab* (ID)*, Fakher Rahim* (ID)*, Bagher Larijani* (ID)*, and Parisa Goodarzi*

© Springer Nature Switzerland AG 2020
B. Arjmand et al. (eds.), *Biomedical Product Development: Bench to Bedside*,
Learning Materials in Biosciences, https://doi.org/10.1007/978-3-030-35626-2_12

12.1 What You Will Learn in This Chapter

Good Clinical Practice (GCP) guidelines share specific features, involve all stakeholders of clinical trials, and clarify their respective responsibilities. Accordingly, in this chapter, all aspects of GCP will be discussed in detail. Before conducting the trial, there are some requirements such as justification of the research – especially ethical wise – and establishment of procedures for documentation. Further, the risk management processes and principles are other essential parts of GCP which will be addressed. Finally, the most important challenges and different ways to overcome them will be introduced.

12.2 Rationale and Importance

GCP guideline is crucial for designing, running, and analyzing clinical trials since it characterizes ethical and scientific rules and standards for each part of the study to protect human safety and rights during the trial. Also, there should be a general and universal framework available for international and national authorities to act as the common language and the criteria for evaluating the GCPs and clinical trials all over the world. Accordingly, GCP guidelines play an important role in providing harmonization among various studies from different regions in order to facilitate research processes. Moreover, GCP guidelines focus on analyzing the process of the trials to improve the quality of the results. In short, GCPs provide doctrines for conducting the trials.

12.3 Features of Good Clinical Practice

The International Conference on Harmonization-GCP as an international standard is used to monitor the design, conduct, and report of clinical trials from ethical and scientific points of view and also defines various aspects of the clinical researches. Consequently, the ICH-GCP contains the following elements to ensure that the trial subjects are protected and the data obtained from clinical testing has sufficient quality and integrity:

- Approval of the protocol- and trial-related requirements by Institutional Review Board (IRB) and/or Independent Ethics Committee (IEC)
- Requirements for safety monitoring
- Voluntary informed consent
- Management of data and archiving requirements
- Clinical trial responsibilities of investigator, sponsor, monitor, and regulatory authorities [1]

The significance of clinical trials in contributing to healthcare system is as conspicuous as the improvement of the lifespan and well-being of humans through developing new medicament and treatment. Also, there is a growing requirement for establishing a dynamic management system to perform clinical studies and promote ethical trials regarding the increasing authority [2].

12.3.1 Institutional Review Board/Independent Ethics Committee (IRB/IEC)

The institutional ethics committee should formulate standard operating procedures (SOPs), which are detailed and written instructions to achieve uniformity of the performance of a particular function. On the other hand, Institutional Review Board (IRB) as an independent authorization gathers some scientific and nonscientific members to protect human rights and safety of trial participants. Further, it approves and provides continuous review of the protocol and its amendments and also monitors the processes used in obtaining informed consent [2, 3].

Independent Ethics Committee (IEC) brings some medical and nonmedical members to – similar to the IRB – protect rights and safety of the trial participants and also provides public assurance of that protection. It reviews and approves desired opinion on the protocol, the investigator's suitability, the facilities, and the methods and material used in documenting informed consent.

It is recommended that the IRB/IEC includes at least five members whom one of them is a nonscientific member and another one is separate from the institutional site. The IRB/IEC should accomplish its functions according to written operating procedures, maintain written records of the meetings, and comply with GCP or the rest of the applicable standards and guidelines [3].

An IRB/IEC should secure well-being, safety, and rights of all trial subjects (especially vulnerable ones). Moreover, it should attain the following documents:
- Trial protocol/amendments
- Written informed consent form(s) and its/their updates and subject recruitment procedures (e.g., advertisements)
- Written information which is provided to subjects
- Investigator's brochure (IB)
- Available safety information
- Data of payments (if any) to the trial subjects and available compensation
- The investigator's resume and/or other documented evidences of qualifications
- Any other documents related to the IRB/IEC responsibilities

In addition, the IRB/IEC should review and inspect the clinical trial and create a record of the reviewed documents and set deadlines for the following items:
- Approval or desired opinion
- Any revisions preceding to its approval/desired opinion
- Disapproval or negative opinion
- Termination or suspension of any previous approval or desired opinion.

Some other responsibilities of the IRB/IEC are listed in ▶ Box 12.1 [1, 3].

> **Box 12.1   Responsibilities of IRB/IEC**
> The RB/IEC should:
> - Consider the qualifications of the investigator for the proposed trial.
> - Conduct continuing review of each ongoing trial at intervals appropriate to the degree of risk to human subjects.

- Determine that the proposed protocol (or other documents) adequately addresses relevant ethical concerns and meets applicable regulatory requirements, when a nontherapeutic trial is to be carried out with the consent of the subject's legally acceptable representative.
- Review both the amount and method of payment to subjects, and ensure that its information is described in the informed consent.
- Retain all relevant records (e.g., written procedures, membership lists, records of meetings) for a period of at least 3 years after completion of the trial (in case of the request from the regulatory authority).

12.3.2 Investigator

The investigator (or the investigators) should be totally educated, trained, and experienced to be responsible for conducting the trial properly. He or she should be adequately capable regarding the GCP and applicable regulatory requirements and provide its evidence. Additionally, he or she should permit sponsor's supervision and inspection by the regulatory authorities. A list of qualified individuals to whom the investigator has assigned tasks should be maintained [3].

Adequate and safe medical care of subjects during the trial and ensuring the maintenance of the suitable medical care and relevant follow-up procedures for a period of time (dependent on the targeted disease, the trial features, and the interventions) are the investigator's responsibilities. ▶ Box 12.2 demonstrates some other responsibilities of the investigator, as summarized from the Guidelines for Good Clinical Practice (GCP) for Trials on Pharmaceutical Products. WHO Technical Report Series, No. 850 [4].

Box 12.2   Responsibilities of the Investigator

The investigator must:
- Assure the unbiased selection and sufficient number of appropriate subjects according to the protocol.
- Agree and sign the protocol with the sponsor, and confirm in writing that he or she will work according to the protocol and GCP.
- Give sufficient information to subjects about the trial, and establish secure safeguards of confidentiality of research data.
- Be completely familiar with the properties, effects, and safety of the investigational new product.
- Give notification of the trial to the drug regulatory authority as governed by national regulations.
- Ensure that the proposed clinical trial has been reviewed and accepted in writing by the relevant independent ethics committees prior to its initiation.
- Notify (documented) the relevant health authorities, the sponsor, and the ethics committee immediately in the case of serious adverse events or reactions.
- Maintain a list of qualified persons to whom he/she has delegated duties.
- Inform the trial subjects, the drug regulatory authority, the ethics committee, and the sponsor if the trial is prematurely terminated.
- Prepare and submit a final report to the drug regulatory authority after completion of the trial.

12.3.3 Sponsor

As it was mentioned in ▶ Chap. 11, the sponsor, who establishes, finances and organizes a clinical trial conduction, can be a pharmaceutical company, independent organization, an individual, or the investigator [4]. Sponsor is responsible for choosing the investigator as well as being assured of the qualification and availability to conduct the trial and taking into account the appropriateness and availability of the trial site and facilities [4].

▶ Box 12.3 depicts some other examples of the sponsor's responsibilities, as summarized from the Guidelines for Good Clinical Practice (GCP) for Trials on Pharmaceutical Products. WHO Technical Report Series, No. 850 [3, 4].

Box 12.3   Responsibilities of the Sponsor

The sponsor should:

- Assure the investigator's agreement to undertake the clinical trial as described in the protocol, and according to GCP, and to accept procedures for data recording, monitoring, audits, and inspections.
- Provide the investigator with available chemical/pharmaceutical, toxicological, pharmacological, and clinical data (including data from previous and ongoing trials) regarding the investigational product.
- Supply the investigational new product.
- Inform the investigator of any immediately relevant information on safety that becomes available during a clinical trial.
- Establish written SOPs to comply with GCP whenever warranted.
- Identify, evaluate, and control risks to processes and data.
- Provide adequate compensation or treatment for subjects in the event of trial-related injury or death as required by national law or regulations.
- Appoint suitable and appropriately trained monitors and clinical research support personnel, and provide ongoing training.
- Establish a system of quality assurance (including independent auditing) to ensure that the coordination of the clinical trial and the generation, documentation, and reporting of data comply with the protocol, GCP standards, and other applicable regulatory requirements.
- Assign special forms for reporting any adverse event that occurs during the clinical trial.
- Notify the investigator, ethics committee, and relevant authorities of the decision about early termination of the trial and its reasons.

12.3.4 **Monitor**

The monitor, assigned by the sponsor, serves as the main communication link between the investigator and the sponsor. Complexity of the trial and types of involved centers direct the number of monitors needed to ensure adequate monitoring.

Not only the monitor should be properly trained and educated but also he or she should know any aspect of the investigational new product and the protocol requirements. Adequate medical, pharmaceutical, clinical trial experience, and scientific qualifications are other expected requirements on the monitor [4].

The monitor controls the trial progress and the compliance of the study with the GCP, protocol, and regulatory requirements, as the main responsibilities. More duties of the monitor can be seen in ▶ Box 12.4, as summarized from the Guidelines for Good Clinical Practice (GCP) for Trials on Pharmaceutical Products. WHO Technical Report Series, No. 850 [4].

> **Box 12.4   Responsibilities of the Monitor**
> The monitor should:
> - Assess the trial site preceding the clinical trial to ensure that the facilities (including laboratories, equipment, and staff) are adequate.
> - Ensure that all staff assisting the investigator in the trial have been adequately informed about and will comply with the details of the trial protocol.
> - Assist the investigator in reporting the data and results of the trial to the sponsor.
> - Make sure that all case report forms (CRFs) are correctly filled out in accordance with original observations (see ▶ Sect. 12.4.2.1).
> - Facilitate communication between the investigator and sponsor, and be available to the investigator at all times for reporting of adverse events or consultation on other trial-related matters.
> - Give assistance to the investigator in notifying the drug regulatory authority of the clinical trial and submitting any necessary documentation.
> - Submit a written monitor report to the sponsor after each site visit and after all relevant telephone calls, letters, and other contacts with the investigator.

12.3.5 Informed Consent

By assigning informed consent, participants confirm their involvement in a trial after getting relevant information about all aspects of that. It is written, signed, and dated which includes explanations of the objectives of the trial, its procedures, the subject's responsibilities, the possible risks and potential and predicted benefits, the expected duration of subject's participation, and the estimated number of them. Further, it addresses that the subject's identity will remain confidential after publishing the results. Additionally, it involves anticipated payments or expenses (if any) to the subjects. The language used in the information should be quite understandable to subjects, and they should know that they have access to appropriate persons to get further information [3].

12.4 Provisions and Prerequisites to Proceed with a Clinical Trial

12.4.1 Justification of Clinical Research

Prior to implementing the clinical trial, some specific conditions and requirements need to be considered. One of the most important considerations is that the specific purposes, risks or benefits, and problems of a clinical trial be completely examined and the chosen options be justified ethically and scientific wise. All researches which involve human subjects are to be conducted in line with the ethical principles of the current version of the Declaration of Helsinki. Evidence of safety and final clinical use of the new product should be thoroughly provided. Also, for planning subsequent trials, information about manufacturing procedures and safety and efficacy of the investigational product attained in the former and current trials is required [4]. In brief, it should be noted that the clinical trial is designed focusing on the best standard of care as Helsinki Declaration states that the effectivity, risks, and benefits of the new investigational treatment should overweigh those of current available methods [5].

Regarding patient recruitment to clinical trials, there are some critical aspects. One of them is the potential expected benefit considered by patients when deciding to take part in the trial. For instance, many patients participate in a study due to their difficulties in accessing expensive treatments and diagnostics needed for their diseases. So, every patient performs own benefit/risk evaluation, and there is the possibility that they participate because they have limited choices [5].

Ethical concerns are another major aspect. Actually, for providing an ethical research, seven requirements can be proposed for evaluation:

1. Social or scientific values: The research should improve health and well-being or increase knowledge.
2. Scientific validity: Accepted scientific principles and methods must be used, and the conduction methodology should be rigorous to generate reliable data.
3. Fair subject selection: It should not target vulnerable individuals and distribute the risks and benefits fairly.
4. Favorable risk-benefit ratio: It should increase the predicted benefits to patients and minimize the potential risks. Also, the possible benefits to the subjects and society outweigh risks.
5. Independent review: It should be reviewed by individuals unaffiliated with the research to minimize potential bias and conflicts.
6. Informed consent: It must provide the information to subjects to ensure that they are enrolled with their authorization.
7. Respect for subjects: The individuals should be respected even when they refuse to participate and during their participation and after it ends [6].

As it was mentioned in the previous subtitle, the investigator and sponsor should reach an agreement on the protocol, SOPs, the monitoring and audit, and the allocation of responsibilities. Obviously, the trial site should be adequate to ensure the safety and efficiency of conducting the trial. There should be regulations governing the way of conducting studies in countries in which clinical trials are performed and all groups involved in a trial should be consistent with current national regulations. If there is no regulation in the country or supplementation is required, relevant government authorities may decide to choose these guidelines to the basis of conducting a trial [4].

12.4.2 Essential Documents for Conducting a Clinical Trial

Essential documents (e.g., informed consent, IB, and protocol) are the documents which independently and cooperatively allow evaluation of conducting a trial and data quality. These documents present the compliance of the sponsor, investigator, and monitor with the GCP standards and regulatory requirements. Filling in them can assist in managing the trial by investigator, sponsor, and monitor successfully. The regulatory authorities inspect the essential documents to approve the validity and cogency of the trial and the integrity of the obtained data. The documents can be divided into three categories: before, during, and after completion of the trial. There are some documents which should be established and recorded before the clinical phase of the trial begins, such as IB, CRF, insurance information, and medical tests. Additionally, other papers including IB and

CRF revisions, monitoring reports, and communications data should be added during the trial to show that all new relevant data is archived as it becomes accessible. After the completion or early termination of the trial, some other forms (like final reports) will be added and be in a file together with previously mentioned documents. Accordingly, tables in Subtitles 8.2, 8.3, and 8.4 of Integrated Addendum to ICH E6 (R1): Guideline for Good Clinical Practice E6(R2) demonstrate these different necessary documents and their purposes in more detail for further information [3].

12.4.2.1 Case Report Form

The CRF is a document intended to record all the required information on trial subjects which is mentioned in the protocol and must be reported to the sponsor [3]. The procedures by which the data is collected guarantee reservation, protection, and recovery of information and permit easy access for evaluation and verification [4].

12.4.2.2 Investigator's Brochure

The IB is a collection of data (clinical and nonclinical) on the investigational new product that is related to its study in human subjects [3]. It provides information about many principal aspects of the protocol such as dose and its frequency, methods and routes of administration, and safety monitoring procedures. It should be presented simply, objectively, and also in a nonpromotional form to enable the investigator to assess the risk and benefit of the suitability of the proposed trial in an unbiased manner. Generally, it is the sponsor's duty to ensure that the investigator has access to an up-to-date IB and also the investigator should provide that to the IRB/IEC. The IB should include title page which consists of items such as the sponsor's name, the identity of investigational products, and the release date. Moreover, it may include a confidentiality statement. Besides, the following sections can be found as contents of the IB:

- Table of contents
- Summary
- Introduction
- Physical, chemical, and pharmaceutical properties and formulation
- Nonclinical studies
- Effects in humans
- Guide and summarized data for the investigator [3].

12.5 Protocol of Clinical Study

A written protocol, which the investigator and sponsor agreed upon, is the basis of the clinical trial. It should be noted that the subsequent changes must be agreed upon and signed and attached to the protocol as amendments. The protocol or any other relevant documentation should address the aim of the trial and why it should be undertaken on humans, the risks, and the means for ensuring that the subjects are informed. Also, it should be reviewed from a scientific and ethical standpoints by review bodies independent of the investigator and sponsor [4]. The contents of a trial protocol include the following topics:

12.5.1 General Information

The general information should include protocol title, identifying number and date, name and address of the sponsor and monitor, name and title of the investigator, and address and telephone number of the trial sites and similar information.

12.5.2 Background and Summary

It consists of the name and description of the investigational products. Findings from relevant clinical trials and significant nonclinical studies should be summarized and addressed. Descriptions of the target population and the route of administration and a statement about conducting the trial in compliance with protocol, GCP, and other regulatory requirements need to be included. Furthermore, a précis of the detected and possible risks and benefits to human subjects is required.

12.5.3 Objectives and Design of the Trial

Objectives and purpose of the trial should be described in detail. The trial design, which provides the trial integrity and the data credibility, should include type of the trial and a flowchart of the trial design, primary and secondary endpoints to be detected, the actions done to prevent bias, treatment, and dosage and regimen of the investigational new products, the expected period of participation, and duration of all study processes and stopping rules.

12.5.4 Inclusion and Exclusion Criteria

This section introduces subject inclusion, exclusion, and withdrawal criteria which some of their points are when and how to withdraw subjects and replace them, the type, and schedule of data which need to be collected for withdrawn subjects.

12.5.5 Treatment and Intervention

Treatments should be described focusing on the names of all products, dosing program, route of administration, and the treatment periods. Also, permitted and not permitted medicines before and during the trial and procedures for monitoring subject compliance need to be determined.

12.5.6 Safety and Efficacy Evaluation

Safety and efficacy parameters and their assessing, recording, and analyzing methods should be specified.

Further, statistics, direct access to source data, quality control and assurance, ethics, data processing and record maintenance, financing and insurance, publication policy, and supplements are noted in ICH-GCP E6(R2) [3].

12.6 Risk Management

Risk management should follow the rules and frameworks of international, national, or regional authorities. However, they may differ regarding to specific situations. The frameworks consist of some tenets such as the responsibilities for running the risk management, definition of the approaches to evaluate the quality of the risk management, and recognition of the relationship between a special phase of the trial and other parts of the process.

Additionally, risk management has different parts including risk assessment (explained in ▶ Chap. 10), risk treatments which aim to decrease the risks by eradicating them or attaining an adequate level, and risk review. Risk review, as a crucial part, has some objectives:

- Testing to confirm the efficacy and effectiveness of the Investigational New Drugs (INDs)
- Revising the plan and design of risk management in order to improve the quality and fulfill the needs [7]

12.7 Challenges and Future Direction

GCP guidelines, protocol, and other requirements pose some challenges and obstacles which are briefly mentioned:

- Time management: Estimating and managing the assessment process timing is of great importance, and the assessment must be done before the medical device (MD) is distributed.
- Acceptability: In order to reduce the execution bias and protocol deviations, an expertise-based randomized clinical trial can be done.
- Blinding: It may be complete or partial, and its assessment should be possible in the procedure.
- Comparison: The comparators should be selected based on standards and evidences related to the investigational treatment. It should be considered that more than one comparator may be needed.
- Learning effect: The learning phase and experience (before and required during the study) must be considered to evaluate any possible relevant benefit.
- Transparency: The results and the protocol of the study should be clearly illustrated and publicly available.
- Regulatory and ethical requirements: There should be available information on regulatory and ethical requirements in MD studies which might be county specific [8, 9].

> **Take-Home Messages**
>
> - GCP guidelines play an important role in providing harmonization for conducting trials all around the world.
> - GCP guideline includes major stakeholders like investigator, sponsor, monitor, regulatory authorities, and IRB/IEC which are responsible for conducting and managing the trial.
> - Before conducting the trial, it must be justified, and different aspects need to be considered. Besides, there are some documents like the investigator's brochure which are required to be prepared before, during, and after the trial.
> - The clinical trial protocol is the basis of clinical trial, and it identifies the aim of the study and some issues like the trial design and assessment of safety and efficacy.
> - Risk management is designed to improve the quality of the trials and fulfill the requirements.
> - Despite all benefits of GCP guidelines, there are still some challenges and obstacles which need to be taken into account.

References

1. Otte A, Maier-Lenz H, Dierckx RA. Good clinical practice: historical background and key aspects. Nucl Med Commun. 2005;26(7):563–74.
2. Narayanan VA, Fernandes S, Ds S, Kumar P, Castelino L. Good clinical practices: an Indian perspective. Res J Pharm Technol. 2018;11:3209.
3. Integrated Addendum to ICH E6(R1): Guideline for Good Clinical Practice E6(R2). 2016. https://www.ich.org/products/guidelines/efficacy/efficacy-single/article/integrated-addendum-good-clinical-practice.html. Accessed 1 Mar 2018.
4. Guidelines for Good Clinical Practice (GCP) for Trials on Pharmaceutical Products. WHO technical report series, no. 850, Annex 3 – WHO Expert Committee on selection and use of essential medicines, sixth report. 1993. https://apps.who.int/medicinedocs/en/d/Jwhozip13e/7.4.html. Accessed 1995.
5. Bosnjak Pasic M, Vidrih B, Sarac H, Pasic H, Vujevic L, Soldo Koruga A, et al. Clinical trials in developing countries – ethical considerations. Psychiatr Danub. 2018;30(3):285–91.
6. Smith-Tyler J. Informed Consent, Confidentiality, and Subject Rights in Clinical Trials. Proceedings of the American Thoracic Society. 2007;4(2):189–93.
7. European Clinical Research Infrastructures Network Integrating Activity. Guideline on risk management for clinical research. Version 1.0. European Commision. 2015. https://ssl2.isped.u-bordeaux2.fr/OPTIMON/docs/citations/2015-02-16-Guideline%20on%20risk%20management%20for%20clinical%20research_1.0x.pdf. Accessed 16 Feb 2015.
8. McCulloch P, Cook JA, Altman DG, Heneghan C, Diener MK. IDEAL framework for surgical innovation 1: the idea and development stages. Br Med J (BMJ). 2013;346:f3012.
9. Sertkaya A, Birkenbach A, Berlind A, Eyraud J. Examination of clinical trial costs and barriers for drug development. U.S. Department of Health and Human Services. 2014. https://aspe.hhs.gov/system/files/pdf/77166/rpt_erg.pdf. Accessed 25 July 2014.

Further Reading

Books

Allport-Settle MJ. Good clinical practice: pharmaceutical, biologics, and medical device regulations and guidance documents concise reference; volume 2, guidance. 1st ed. Willow Springs: PharmaLogika; 2010.
McGraw MJ, George AN, Shearn SP, Thomas F, Haws TF Jr, Hall RL. Principles of good clinical practice. 1st ed. London/Chicago: Pharmaceutical Press; 2010.

12

Online Resources

Clinical investigation of medical devices for human subjects—good clinical practice: https://www.iso.org/standard/45557.html.

Handbook for good clinical research practice.: https://www.who.int/medicines/areas/quality_safety/safety_efficacy/gcp1.pdf.

Integrated Addendum to ICH E6(R1): Guideline for Good Clinical Practice E6(R2). 2016.: https://www.ich.org/products/guidelines/efficacy/efficacy-single/article/integrated-addendum-good-clinical-practice.html.

WHO-GCP guideline: https://apps.who.int/medicinedocs/en/d/Jwhozip13e/7.4.html.

Ethical Considerations of Biomedical Product Development

Hamideh Moosapour, Asal Zarvani, Maryam Moayerzadeh, and Bagher Larijani

© Springer Nature Switzerland AG 2020
B. Arjmand et al. (eds.), *Biomedical Product Development: Bench to Bedside*,
Learning Materials in Biosciences, https://doi.org/10.1007/978-3-030-35626-2_13

13.1 What You Will Learn in This Chapter

- The importance of research ethics
- The key terms and concepts relating to the research ethics particularly in biomedicine through a conceptual framework
- The history of research ethics
- The origin, structure, and function of research ethics committee
- The ethical considerations concerning the use of animals in biomedical research
- The ethical considerations about human dignity and right in biomedical research in terms of informed consent, privacy, confidentiality, and use of placebo
- The ethical considerations focusing on vulnerable groups

13.2 Rationale and Importance

The importance of ethics can be perceived considering the immorality in terrible history of several projects in the field of biomedical research (e.g., Tuskegee, Willowbrook, Milgram, Stanford Prison, etc.). Tuskegee Syphilis Study, for instance, is conducted by the U.S. Public Health Service (1932–1972). In this study, 400 subjects out of 600 African-American males from a low social economic population were infected with syphilis and observed for 40 years. Free medical check-up was given; however, participants were not aware of their disease. Even though a proven treatment such as penicillin was available in the 1950s, the study continued until 1972 and subjects did not receive treatment. In some cases, when other physicians diagnosed that the participants had syphilis, researchers intervened to stop the treatment. During the research study, many participants died of syphilis. Finally, the study ended in 1973 by the US Department of Health, Education, and Welfare after disclosing its details and a political embarrassment. In 1997, President Clinton under mounting pressure apologized to the study subjects and their families. Because of the advertising from the Tuskegee Syphilis Study, the National Research Act of 1974 was legislated in the United States [1].

How studies such as Tuskegee had been morally justified for physicians and researchers and continued for four decades? How they had not found intuitionally unfairness in such cases?

Although the value of modern medicine rests mainly on scientific research related to the causes of disease, innovative therapies, and so forth, biomedical research is considered to be an area of ethical risk particularly associated with exposing patients or healthy volunteers to a serious harm. In addition, participants demand autonomy to decide on what procedures they are ready to accept more than past [2]. Moreover, communities and experts became more sensitive to the possible effects of biomedical research on the environment, next generations, and animals involved in those researches, as well as the reliability and credibility of research findings. Consequently, this atmosphere has increased importance of strict regulations designed to guarantee that studies on humans and animals keep up ethical and scientific standards. In this chapter, we have reviewed the most prominent concepts, principles, and concerns of ethical reflection on biomedical research.

13.3 Conceptual Framework of Research Ethics in Biomedicine

Research ethics as the generic concept governs the standards of conduct for scientific researchers. Research ethics was first and foremost developed as a concept in medical research; it has been extended to other fields such as social sciences, information technology, engineering, and so forth. Research ethics be mainly discussed in basic principles such as minimizing the risk of harm, obtaining informed consent, protecting anonymity and confidentiality, avoiding deceptive practices, and providing the right to withdraw [3, 4]. Research ethics, also, is distinguished from publication ethics. Whereas research ethics focuses on standards protecting the right of human participants or animals involved in research, publication ethics focuses on standards ensuring public trust in scientific findings, high-quality scientific publications, and people who receive credit for their ideas. In other words, publication ethics are standards to guarantee the research integrity (scientific integrity, or academic integrity). Accordingly, scientific integrity ensures values and practices such as objectivity, clarity, reproducibility, maintenance of academic standards, honesty and rigor in academic publishing, and utility. The violation of scientific integrity is defined as scientific misconduct. Scientific misconduct includes disvalues and malpractices in professional scientific research or academic area such as fabrication, bias, plagiarism, falsification, censorship, inadequate procedural, outside interference, and information security. All participants in academic research area, for example, authors, editors, reviewers of journals, research institutions, and even uninvolved scientific colleagues, are responsible for research and publication ethics [5–11].

13.4 Philosophy of Research Ethics in Biomedicine

Ethics deals with moral duties, principles, values, virtues, and the idea of right and fairness without any harm. It also deals with fundamental questions such as "What is the right action?", "How should I live?", "What kind of person should I be?", "What is justice?", and "What is fairness?". All voluntary and nonmandatory (not commanded by authority) actions of a human being fall within the broad scope of ethical reflections. Ethics as a discipline belongs originally to the subdiscipline of the moral philosophy. Normative ethics seeks the way of reasoning that can be applied to a wide variety of situations. On the other hand, biomedical research ethics is assumed as a branch of applied ethics. Therefore, a rich and complex history of methods developed in normative ethics has been used for ethical reasoning in biomedicine [12]. The theory of four principles is the most well-known method for ethical reflections in biomedical research ethics. It includes the principles of "respect for autonomy," "beneficence," "non-maleficence," and "justice." This theory, developed by two American philosophers (Beauchamp and Childress), is a method for systematic reflection on moral principles and their relevance for determining an optimal course of action in biomedicine by weighing and balancing the claim of these competing principles whenever they are applied. Respect for autonomy is based on the right of individuals to develop their own life plans and make their own choices. Non-maleficence necessitates avoiding harm to patients or their best interests. Beneficence calls us for doing goods that led to further patient's interest, and finally, justice mainly demands that medical goods and services need to be distributed fairly [13]. The key concepts of research ethics like informed consent or rules such as minimizing the risk of harm (as mentioned below) have been originated from those principles. To be specific, informed consent has originated from respect for autonomy and minimizing the risk of harm from non-maleficence [14].

According to a study by Avashti et al., [15] principles of ethics in medical research which are derived from the theory of four principles can be summarized in the following points:

1. Essentiality of research
2. Voluntariness, informed consent, and community agreement
3. Nonexploitation
4. Privacy and confidentiality
5. Risk minimization and precaution
6. Professional competence
7. Accountability and transparency
8. Maximization of the public interest and distributive justice
9. Public domain
10. Totality of responsibility

13.5 History of Research Ethics

A well-known chapter in the history of research with human subjects opened in 1946 when an American military tribunal commenced criminal proceedings against several German physicians and managers for handling medical experiments on thousands of prisoners without getting their consent. Most of the participants of these experiments were permanently crippled or died because of those experiments. Subsequently, in 1948, the Nuremberg Code was established, stating that participants have voluntary consent and also the advantages of research must outweigh the risks. This Code was the first international document which supported voluntary participation and informed consent (See Appendix 1) [16]. The World Medical Association, in 1964, developed guidance for medical practitioners who involved in studies on human participants (the Declaration of Helsinki) [17–19] to govern research ethics and provide some regulations for investigational healthcare and nontherapeutic studies. Subsequently, this approach led to the basis of good clinical practices (GCPs), which is discussed in ▶ Chap. 11 comprehensively [20]. Moreover, it was revised in 1975 as introducing the notion of research ethics committee (REC), as discussed in the next section. Revisions continued until the last revision in 2013, by 64th meeting of the World Medical Association (WMA) [17] (See Appendix 2). Subsequently, the National Commission for the Protection of Human Subjects of Biomedical and Behavioral Research provides the next set of research ethics guidelines in the Belmont Report of 1979. The primary purpose of Belmont report is to protect participants in clinical trials or research studies and outlines:

1. Ethical principles and issues for performing research on human participants
2. Boundaries between medical practices and researches
3. The ideas of respect for beneficence, persons, and justice
4. The use of these principles in evaluating subject selection (justice), informed consent (respect for persons), and risks and benefits (beneficence) [21].

13.6 Research Ethics Committee

REC is an institution for ensuring that proposed studies obey internationally and locally accepted ethical guidelines. Terms such as independent ethics committee (IEC) or institutional review board (IRB) are interchangeably used for REC. RECs have the right to accept, refuse, or withdraw studies or if necessary modify research protocols. All research projects especially experimental ones must receive the official permission of the ethical committee before starting the study. It is completely unethical to enroll any patient before getting officially authorized. RECs also control studies once they have started and, where relevant, take part in follow-up action and surveillance after the end of the research which should be held on a regular period of time. Furthermore, they may set some policies or offer options on ongoing ethical issues in research. The basic logic behind the institution of REC is to protect participants' rights and prefer their interests through a process of team decision-making. Therefore, RECs need a combination of impartial members with proper knowledge and experience for scientific and ethical judgment on research projects. They have to be a combination of experts and non-experts in biomedicine (as a representative of society) and both female and male members. Also, it is required that at least one member would not be affiliated to the institution submitting the project. The list of committee members and their full identity must be clear and publicly available. The main functions of RECs include identifying and weighing up the risks and potential benefits of research, evaluating the process and materials (printed documents and other exact details) that will be used for seeking participants, complete identity of the researchers, budget provider of the study, informed consent which means assessing the recruitment process and any incentives that will be given to participants, confidentiality and the adequacy of confidentiality protections and predicted strategies, assessment of any effects of the study on treating disease or causing any harm, and compensation agreement [2, 11, 22].

13.7 Research Ethics in the Use of Animals in Biomedicine

Animal research is an essential step in biomedical research process because it is usually the first step toward research involving new medical treatments and pharmaceuticals intended for human use. The number of animals used in research has been estimated at 17–70 million animals per year generally, but other sources like Hendee, Fox, and Orlans estimate somewhere between 20 and 35 million. However, using animals for biomedical research is one of the most debated issues [23]. The proponents' utilitarian argument is based on the claim that promoting human health and welfare as the ends justifies animal research as the means. Opponents offer two types of critiques, moral and scientific or technical ones. The former argue the ends cannot justify the means by using two concepts: ethical status and animal rights. The latter argue that the means are not effective to that ends. The moral critiques begun from the nineteenth century as Jeremey Bentham and John Stuart Mill argued that animals deserve moral consideration because they can suffer. Then, the notions of animal's welfare and animal rights formed bases for a modern movement gained strength in the twentieth century. The technical critiques resulted in the proposed principles of replacement, reduction, refinement, relevance, and redundancy avoidance (described in ▶ Chap. 5) in animal researches [11].

To ensure that research on animals is conducted ethically and responsibly, associations and guidelines for the protection of animal rights are developed (see Appendix 3 which represents guidelines developed by the American Psychological Association (APA) for use by psychologists working with nonhuman animals) [24]. Ethical considerations of animal research in this guideline have been classified in ten categories, including general, personnel, facility, care and housing of animals, justification of the research, experimental design, experimental procedures, animal acquisition, field research, educational benefits of the use of animals, and animal disposition.

13.8 Research Ethics Involving Human Experiments

Ethical issues in research on human participants focus on various topics from voluntary participation in research to fair selection and justice. This variety makes these ethical issues a challenge but important charge. Among them, some of the relevant and important topics regarding ethical concerns in clinical studies for the development of biomedical products including informed consent, privacy and confidentiality, and the use of placebo in clinical trials will be discussed in the following.

13.8.1 Informed Consent

Informed consent is a fundamental and thoughtful process (not merely a form) with respect to the autonomy of participants to ensure that their enrollment in the study and continuing participation in studies is an informed and voluntary act. It usually must be explicit, specific, and documented by using a written, signed, and dated form (not implicit, general, or verbal). It also includes study information and focuses on comprehension and voluntary participation. Study information means that it is crucial to disclose all the information, possible risks regardless of the effects on participants' willingness to participate or not, the purpose of the study, benefits, and any available alternative diagnostic or therapeutic options and finally give them the opportunity to ask any questions. In addition, comprehension refers to the mental capacity of participants and level of their understanding of the study, its purpose and benefits, as well as their ability to cooperate with the researchers during the study.

Finally, voluntary participations refer to participant's exercise of free will in deciding whether to take part in research activity. Participants must be neither coerced nor improperly pressured from any participant. For fully voluntary participations, it is required to prevent participants from feeling deceived, persuaded, exploited, shamed, abused, coerced, or otherwise wronged by researchers. As a result, research on human subjects where they do not have the capacity to consent (such as children, patients need to get an emergency treatment, and critically ill, mentally impaired, demented, or unconscious patients) or do not have the rights to refuse (in spite of having the capacity to consent) (e.g., soldiers, institutionalized persons, and prisoners) can be challenging [25–28].

13.8.2 Privacy and Confidentiality

Privacy and confidentiality are two concepts often mistaken to be the same thing. However, they are referred to related but not identical concepts, particularly from an ethical or legal standpoint. Privacy concerns people, whereas confidentiality concerns data. Privacy refers to people's desire to control the access of others to personal matters and personal information (e.g., the access of a researcher to their body for physical examinations or to their personal information). Confidentiality refers to the duty of anyone entrusted with information to keep that information private (e.g., duty of a researcher to keep participants' information private and not tell others including co-workers or family without permission). Since the information gathered from people in biomedical studies has the potential to be embarrassing, harmful, or damaging, it is necessary that research proposals outline strategies to protect privacy and maintain confidentiality [11].

13.8.3 Use of Placebo in Clinical Trials

A placebo is an inert treatment that does not have an active drug ingredient but still has a therapeutic effect (placebo effect). Further, the most studies have shown a strong placebo effect. The effectiveness of these kinds of drugs can be seen even when people are told they are being given a placebo (although their effect is more when people are not told).

In clinical trials, researchers use placebos to distinguish whether a new drug works better compared to an inactive alternative. The most reliable clinical trials use a "double-blinded" approach. In this strategy, neither participants nor the person giving them the drug knows that they may be given a placebo or a drug. At first sight, there are not any ethical issues with applying placebos in clinical trials, assuming participants have been properly informed and given consent, but the ethics of placebo in clinical practice is rather intricate in some cases.

> **Take-Home Messages**
>
> - The standards of research ethics mainly focus on the protection of human participants and animals in health-related researches.
> - Publication ethics should guarantee the high quality of scientific dissemination and research integrity.
> - Scientific misconduct is the contravention of scientific integrity which includes malpractices in professional scientific research or academic area.
> - The four-principle approach in biomedical ethics includes non-maleficence, beneficence, respect for autonomy, and justice.
> - The main aim of Belmont Report which was conducted in 1979 is protecting participants in clinical trials or research studies.
> - Research ethics committee (REC) is an institution for reviewing research protocols to ensure proposed studies obey ethical guidelines and identify the potential risks and benefits of research.
> - Research ethics helps scientists to minimize the risk of harm to human subjects and animals.
> - Guidelines for protecting the animal rights should be developed based on avoiding unnecessary suffering on animals and also controlling the use of animals in research.
> - The principles of ethical research involving human subjects cover all aspects of a clinical study from voluntary participation in research to fair selection and justice.

13

- Informed consent is an essential and thoughtful process in accordance with respect to human rights. It also guarantees that their participation in research is informed and voluntary.
- Researchers are responsible for respecting participant's privacy and also keeping their information confidentially.
- The use of placebo should not make the patient deprived of an effective treatment.
- Vulnerable groups should be involved in research only if the study is for the benefit of them.
- It is essential to consider the ethical aspects of using excess tissue or blood of patients for research purposes.
- The more complexity of research leads to more challenges.

The use of placebo is ethically provided that participating in a study does not make the patient deprived of an effective treatment, according to Helsinki Declaration articles: Every patient even those of a control group in any medical study should be assured the best proven diagnostic and therapeutic approach is being used for them and this does not eliminate the use of inert placebo in studies where no proven diagnostic or therapeutic approach exists [2, 17–20].

13.9 Research Ethics on Vulnerable Groups

Sometimes, people are not able to provide informed consent. These groups named as vulnerable groups such as neonates, infants, pregnant women, children in the age group with a relative understanding of the subject, people with learning disabilities, patients with severe psychiatric disorders and prisoners. Therefore, special ethical concerns exist in research on these cases, in addition to the general concerns of medical research ethics. While the ethical considerations for each of the mentioned groups can be different, a common principle is that these people can participate in a research if only the researcher and REC assure that the research cannot be done on an adult group in other healthy ways because it is needed to design the study for the benefit of them. For instance, research on children is inevitable for several reasons including children have their own disease, the appropriate dose of a drug in children is not the same as those in adults, and children can be more sensitive to adverse effects of a drug [2, 11, 29].

13.10 Research Ethics in Special Areas

There are some distinct areas of experimental research with special ethical considerations which include research on excess blood and other tissues of patients, research on genetics and gene therapy, and research on embryo and stem cell. For example, while using the excess amounts of blood and other tissues for research is primarily sound, it is essential to consider the ethical aspects of this process. In every practice, taking a few more drops of blood because of the probability of additional or repeated tests or removing extra tissue surrounding a tumor as a precaution is usual. However, if excess tissue or blood is provided only for research purposes, it is obvious that consent must be obtained from the patient. Furthermore, sometimes these tissues may be used in the future for special purposes that patients may oppose them. In this case, it is important to obtain informed consent from the patient in these situations. For example, if the placenta may be used for the development of the new abortion drug in the future, the written consent of the patient is mandatory [2].

13.11 Challenges and Future Perspectives

This chapter focuses on major ethical issues in biomedical researches. Nowadays, complex fields of science such as OMICs and biobanking and also the wide spectrum of precision medicine investigations have opened a new horizon with its particular ethical considerations. Accordingly, based on the huge data extracted from the mentioned novel and complex fields, protecting privacy and confidentiality have become more challenging [30–32].

Appendix 1. The Nuremberg Code (1947)

1. The voluntary consent of the human subject is absolutely essential. This means that the person involved should have legal capacity to give consent, should be so situated as to be able to exercise free power of choice, without the intervention of any element of force, fraud, deceit, duress, overreaching, or other ulterior form of constraint or coercion; and should have sufficient knowledge and comprehension of the elements of the subject matter as to enable him to make an understanding and enlightened decision. This latter element requires that before the acceptance of an affirmative decision by the experimental subject there should be made known to him the nature, duration, and purpose of the experiment; the method and means by which it is to be conducted; all inconveniences and hazards reasonably to be expected; and the effects upon his health or person which may possibly come from his participation in the experiment. The duty and responsibility for ascertaining the quality of the consent rests upon each individual who initiates, directs, or engages in the experiment. It is a personal duty and responsibility which may not be delegated to another with impunity.
2. The experiment should be such as to yield fruitful results for the good of society, unprocurable by other methods or means of study, and not random and unnecessary in nature.
3. The experiment should be so designed and based on the results of animal experimentation and a knowledge of the natural history of the disease or other problem under study that the anticipated results justify the performance of the experiment.
4. The experiment should be so conducted as to avoid all unnecessary physical and mental suffering and injury.
5. No experiment should be conducted where there is an *a priori* reason to believe that death or disabling injury will occur; except, perhaps, in those experiments where the experimental physicians also serve as subjects.
6. The degree of risk to be taken should never exceed that determined by humanitarian importance of the problem to be solved by the experiment.

7. Proper preparations should be made and adequate facilities provided to protect the experimental subject against even remote possibilities of injury, disability or death.
8. The experiment should be conducted only by scientifically qualified persons. The highest degree of skill and care should be required through all stages of the experiment of those who conduct and engage in the experiment.
9. During the course of the experiment the human subject should be at liberty to bring the experiment to an end if he has reached the physical or mental state where continuation of the experiment seems to him to be impossible.
10. During the course of the experiment the scientist in charge must be prepared to terminate the experiment at any stage, if he has probable cause to believe, in the exercise of good faith, superior skill and careful judgment required of him, that a continuation of the experiment is likely to result in injury, disability, or death to the experimental subject.

Appendix 2. Declaration of Helsinki (1975)

Adopted by the 18th WMA General Assembly Helsinki, Finland, June 1964 and amended by the 29th WMA General Assembly, Tokyo, Japan October 1975.

Introduction

It is the mission of the medical doctor to safeguard the health of the people. His or her knowledge and conscience are dedicated to the fulfilment of this mission.

The Declaration of Geneva of the World Medical Association binds the doctor with the words: 'The health of my patient will be my first consideration', and the International Code of Medical Ethics declares that 'Any act or advice which could weaken physical or mental resistance of a human being may be used only in his interest'.

The purpose of biomedical research involving human subjects must be to improve diagnostic, therapeutic and prophylactic procedures and the understanding of the etiology and pathogenesis of disease.

In current medical practice, most diagnostic, therapeutic or prophylactic procedures involve hazards. This applies *a fortiori* to biomedical research.

Medical progress is based on research which ultimately must rest in part on experimentation involving human subjects. In the field of biomedical research, a fundamental distinction must be recognized between medical research in which the aim is essentially diagnostic or therapeutic for a patient, and medical research the essential object of which is purely scientific and without direct diagnostic or therapeutic value to the person subjected to the research.

Special caution must be exercised in the conduct of research which may affect the environment, and the welfare of animals used for research purposes must be respected.

Because it is essential that the results of laboratory experiments be applied to human beings to further scientific knowledge and to help suffering humanity, the World Medical Association has prepared the following recommendations as a guide to every doctor in biomedical research involving human subjects. They should be kept under review in the future. It must be stressed that the standards as drafted are only a guide to physicians all over the world. Doctors are not relieved from criminal, civil and ethical responsibilities under the laws of their own countries.

I Basic principles

1. Biomedical research involving human subjects must conform to generally accepted scientific principles and should be based on adequately performed laboratory and animal experimentation and on a thorough knowledge of the scientific tradition.
2. The design and performance of each experimental procedure involving human subjects should be clearly formulated in an experimental protocol which should be transmitted to a specially appointed independent committee for consideration, comment, and guidance.
3. Biomedical research involving human subjects should be conducted only by scientifically qualified persons and under the supervision of a clinically competent medical person. The responsibility for the human subject must always rest with a medically qualified person and never rest on the subject of the research, even though the subject has given her consent.
4. Biomedical research involving human subjects cannot legitimately be carried out unless the importance of the objective is in proportion to the inherent risk to the subject.
5. Every biomedical research project involving human subjects should be preceded by careful assessment of predictable risks in comparison with foreseeable benefits to the subject or to others. Concern for the interests of the subject must always prevail over the interest of science and society.
6. The right of the research subject to safeguard his or her integrity must always be respected. Every precaution should be taken to respect the privacy of the subject and to minimize the impact of the study on the subject's physical and mental integrity and on the personality of the subject.
7. Doctors should abstain from engaging in research projects involving human subjects unless they are satisfied that the hazards involved are believed to be predictable. Doctors should cease any investigation if the hazards are found to outweigh the potential benefits.
8. In publication of the results of his or her research, the doctor is obliged to preserve the accuracy of the results. Reports of experimentation not in accordance with the principles laid down in this declaration should not be accepted for publication.
9. In any research on human beings, each potential subject must be adequately informed of the aims, methods, anticipated benefits and potential hazards of the study and the discomfort it may entail. He or she should be informed that he or she is at liberty to abstain from participation in the study and that he or she is free to withdraw his or her consent to participation at any time. The doctor should then obtain the subject's freely given informed consent, preferably in writing.
10. When obtaining informed consent for the research project, the doctor should be particularly cautious if the subject is in a dependent relationship to him or her or may consent under duress. In that case, informed consent should be obtained by a doctor who is not engaged in the investigation and who is completely independent of this official relationship.

11. In cases of legal incompetence, informed consent should be obtained from the legal guardian in accordance with national legislation. Where physical or mental incapacity makes it impossible to obtain informed consent, or when the subject is a minor, permission from the responsible relative replaces that of the subject in accordance with the national legislation.
12. The research protocol should always contain a statement of ethical consideration involved and should indicate that the principles enunciated in the present declaration are complied with.

II Medical research combined with professional care (clinical research)

1. In the treatment of the sick person, the doctor must be free to use a new diagnostic and therapeutic measure, if in his or her judgment it offers the hope of saving life, re-establishing health or alleviating suffering.
2. The potential benefits, hazards and discomfort of a new method should be weighed against the advantages of the best current diagnostic and therapeutic methods.
3. In any medical study, every patient – including those of a control group, if any – should be assured of the best proven diagnostic and therapeutic method.
4. The refusal of the patient to participate in a study must never interfere with the doctor–patient relationship.
5. If the doctor considers it essential not to obtain informed consent, the specific reasons for this proposal should be stated in the experimental protocol for transmission to the independent committee.
6. The doctor can combine medical research with professional care, the objective being the acquisition of new medical knowledge, only to the extent that medical research is justified by its potential diagnostic or therapeutic value for the patient.

III Non-therapeutic biomedical research involving human subjects (non-clinical biomedical research)

1. In the purely scientific application of medical research carried out on a human being, it is the duty of the doctor to remain the protector of the life and health of that person on whom biomedical research is carried out.
2. The subjects should be volunteers – either healthy persons or patients for whom the experimental design is not related to the patient's illness.
3. The investigator or the investigating team should discontinue the research if in his/her or their judgment it may, if continued, be harmful to the individual.
4. In research on man, the interest of science and society should never take precedence over considerations related to the well-being of the subject.

Appendix 3. Guidelines for Ethical Conduct in the Care and Use of Animals

Developed by the American Psychological Association's Committee on Animal Research and Ethics (CARE)[1]

I. General

A. In the ordinary course of events, the acquisition, care, housing, use, and disposition of animals should be in compliance with relevant federal, state, local, and institutional laws and regulations and with international conventions to which the United States is a party. In accordance with Principle 3(d) of the Ethical Principles of Psychologists of APA, when federal, state, provincial, organizational, or institutional laws, regulations, or practices are in conflict with Association Guidelines, psychologists should make known their commitment to Association Guidelines and, whenever possible, work toward resolution of the conflict.

B. Psychologists and students working with animals should be familiar with these Guidelines, which should be conspicuously posted in every laboratory, teaching facility, or other setting in which animals are maintained and used by psychologists and their students.

C. Violations of these Guidelines should be reported to the facility supervisor whose name is appended at the end of this document. If not resolved at the local level, allegations of violations of these Guidelines should be referred to the APA Committee on Ethics, which is empowered to impose sanctions. No psychologists should take action of any kind against individuals making, in good faith, a report of violation of these Guidelines.

D. Individuals with questions concerning these Guidelines should consult with the Committee on Animal Research and Experimentation.

E. Psychologists are strongly encouraged to become familiar with the ethical principles of animal research. To facilitate this, the Committee on Animal Research and Experimentation will maintain a list of appropriate references.

II. Personnel

A. A supervisor, experienced in the care and use of laboratory animals, should closely monitor the health, comfort, and humane treatment of all animals within the particular facility.

B. Psychologists should ensure that personnel involved in their research with animals be familiar with these Guidelines.

C. It is the responsibility of the supervisor of the facility to ensure that records of the accession, utilization, and disposition of animals are maintained.

---

1   The following text is quoted exactly from: American Psychological Association. Guidelines for ethical conduct in the care and use of animals. Journal of the Experimental Analysis of Behavior. 1986 Mar;45(2):127.

D. A veterinarian should be available for consultation regarding: housing, nutrition, animal-care procedures, health, and medical attention. The veterinarian should conduct inspections of the facility at least twice a year.
E. Psychologists should ensure that all individuals who use animals under their supervision receive explicit instruction in experimental methods and in the care, maintenance, and handling of the species being studied. Responsibilities and activities of all individuals dealing with animals should be consistent with their respective competencies, training and experience in either the laboratory or the field setting.
F. It is the responsibility of the psychologist to ensure that appropriate records are kept of procedures with animals.
G. It is the responsibility of the psychologist to be cognizant of all federal, state, local, and institutional laws and regulations pertaining to the acquisition, care, use, and disposal of animals. Psychologists should also be fully familiar with the NIH Guide for the Care and Use of Laboratory Animals.

III. Facilities

A. The facilities housing animals should be designed to conform to specifications in the NIH Guide for the Care and Use of Laboratory Animals.
B. Psychologists are encouraged to work toward upgrading the facilities in which their animals are housed.
C. Procedures carried out on animals 128 APA GUIDELINES are to be reviewed by a local institutional animal care and use committee to ensure that the procedures are appropriate and humane. The committee should have representation from within the institution and from the local community. If no representative from the local community is willing to serve, there should be at least one representative on the committee from a non-science department. In the event that it is not possible to constitute an appropriate local institutional animal care and use committee, psychologists should submit their proposals to the corresponding committee of a cooperative institution.

IV. Acquisition of Animals

A. When appropriate, animals intended for use in the laboratory should be bred for that purpose.
B. Animals not bred in the psychologist's facility are to be acquired lawfully. The U.S. Department of Agriculture (USDA) may be consulted for information regarding suppliers.
C. Psychologists should make every effort to ensure that those responsible for transporting the animals to the facility provide adequate food, water, ventilation, and space, and impose no unnecessary stress upon the animals.
D. Animals taken from the wild should be trapped in a humane manner.
E. Endangered species or taxa should be utilized only with full attention to required permits and ethical concerns. Information can be obtained from the Office of Endangered Species, U.S. Department of the Interior, Fish and Wildlife Service, Washington, D.C. 20240. Similar caution should be used in work with threatened species or taxa.

V. Care and Housing of Animals

Responsibility for the conditions under which animals are kept, both within and outside of the context of active experimentation or teaching, rests jointly upon the psychologist and those individuals appointed by the institution to administer animal care. Animals should be provided with humane care and healthful conditions during their stay in the facility. Psychologists are encouraged to consider enriching the environments of their laboratory animals, where appropriate.

VI. Justification of the Research

A. Research should be undertaken with a clear scientific purpose. There should be a reasonable expectation that the research will a) increase knowledge of the processes underlying the evolution, development, maintenance, alteration, control, or biological significance of behavior, b) increase understanding of the species under study, or c) provide results that benefit the health or welfare of humans or other animals.

B. The scientific purpose of the research should be of sufficient potential significance as to outweigh any harm or distress to the animals used. In this regard, psychologists should act on the assumption that procedures that would produce pain in humans will also do so in other animals.

C. The psychologist should always consider the possibility of using alternatives to animals in research and should be familiar with the appropriate literature.

D. Research on animals may not be conducted until the protocol has been reviewed by the institutional animal care and use committee to ensure that the procedures are appropriate and humane.

E. The psychologist should monitor the research and the animals' welfare throughout the course of an investigation to ensure continued justification for the research.

VII. Experimental Design

Humane considerations should constitute one of the major sets of factors that enter into the design of research. Two particularly relevant considerations should be noted:

1. The species chosen for study should be well-suited to answer the question(s) posed. When the 129 APA GUIDELINES research paradigm permits a choice among species, the psychologist should employ that species which appears likely to suffer least.

2. The number of animals utilized in a study should be sufficient to provide a clear answer to the question(s) posed. Care should be exercised to use the minimum number of animals consistent with sound experimental design, especially where the procedures might cause pain or discomfort to the animals.

VIII. Experimental Procedures

Humane consideration for the wellbeing of the animal should be incorporated into the design and conduct of all procedures involving animals. The conduct of all procedures is governed by Guideline VI.

A. Procedures which involve no pain or distress to the animal, or in which the animal is anesthetized and insensitive to pain throughout the procedure and is euthanized before regaining consciousness, are generally acceptable.

B. Procedures involving more than momentary or slight pain not relieved by medication or other acceptable methods, should be undertaken only when the objectives of the research cannot be achieved by other methods.

C. Procedures involving severe distress or pain that is not alleviated require strong justification. An animal observed to be in a state of severe distress or chronic pain that cannot be alleviated and that is not essential to the purposes of the research, should be euthanized immediately.

D. When aversive or appetitive procedures appear to be equivalent for the purposes of the research, then appetitive procedures should be used. When using aversive stimuli, psychologists should adjust the parameters of stimulation to levels that appear minimal, though compatible with the aims of the research. Psychologists are encouraged to test painful stimuli on themselves whenever reasonable. Whenever consistent with the goals of the research, consideration should be given to providing the animal with control of painful stimulation.

E. Procedures involving extensive food or water deprivation should be used only when minimal deprivation procedures are inappropriate to the design and purpose of the research.

F. Prolonged physical restraint should be used only if less stressful procedures are inadequate to the purposes of the study. Convenience to the psychologist is not a justification for prolonged restraint.

G. Procedures that entail extreme environmental conditions, such as high or low temperatures, high humidity, modified atmospheric pressure, etc. should be undertaken only with particularly strong justification.

H. Studies entailing experimentally induced prey killing or intensive aggressive interactions among animals should be fully justified and conducted in a manner that minimizes the extent and duration of pain.

I. Procedures entailing the deliberate infliction of trauma should be restricted and used only with very strong justification. Whenever possible, without defeating the goals of the research, animals used in such research should be anesthetized.

J. Procedures involving the use of paralytic agents without reduction in pain sensation require particular prudence and humane concern. Utilization of muscle relaxants or paralytics alone during surgery, without general anesthesia, is unacceptable, and shall not be used.

K. Surgical procedures, because of their intrusive nature, require close supervision and attention to humane considerations by the psychologist.

   1. All surgical procedures and anesthetization should be conducted under the direct supervision of a scientist who is competent in the use of the procedure. 130 APA GUIDELINES

2. If the surgical procedure is likely to cause greater discomfort than that attending anesthetization, and unless there is specific justification for acting otherwise, animals should be maintained under anesthesia until the procedure is ended.
3. Sound post-operative monitoring and care should be provided to minimize discomfort, and to prevent infection and other untoward consequences of the procedure.
4. As a general rule, animals should not be subjected to successive surgical procedures unless these are required by the nature of the research, the nature of the surgery, or for the well-being of the animal. However, there may be occasions when it is preferable to carry out more than one procedure on a few animals rather than to carry out a single procedure on many animals. For instance, there may be experimental protocols where it would be appropriate to carry out acute terminal surgical procedures on animals scheduled for euthanasia as part of another protocol rather than to utilize additional animals.

IX. Field Research

A. Psychologists conducting field research should disturb their populations as little as possible. Every effort should be made to minimize potential harmful effects of the study on the population and on other plant and animal species in the area.
B. Research conducted in populated areas should be done with respect for the property and privacy of the inhabitants of the area.
C. Particular justification is required for the study of endangered species. Such research should not be conducted unless all requisite permits are obtained.

X. Educational Use of Animals

A. For educational purposes, as for research purposes, consideration should always be given to the possibility of using non-animal alternatives. When animals are used solely for educational rather than research purposes, the consideration of possible benefits accruing from their use vs. the cost in terms of animal distress should take into account the fact that some procedures which can be justified for research purposes cannot be justified for educational purposes. Similarly, certain procedures, appropriate in advanced courses, may not be appropriate in introductory courses.
B. Classroom demonstrations involving animals should be used only when instructional objectives cannot effectively be achieved through the use of videotapes, films or other alternatives. Careful consideration should be given to the question of whether the type of demonstration is warranted by the anticipated instructional gains.
C. Animals should be used for educational purposes only after review by a departmental committee or by the local institutional animal care and use committee.
D. Psychologists are encouraged to include instruction and discussion of the ethics and values of animal research in courses, both introductory and advanced, which involve or discuss the use of animals.
E. Student projects involving pain or distress to animals should be undertaken judiciously and only when the training objectives cannot be achieved in any other way.

F. Demonstrations of scientific knowledge in such contexts as exhibits, conferences, or seminars do not justify the use of painful procedures or surgical interventions. Audio-visual alternatives should be considered.

XI. Disposition of Animals

A. When the use of an animal is no longer required by an experimental protocol or procedure, alternatives to euthanasia should be considered.
   1. Animals may be distributed to colleagues who can utilize them. 131 APA GUIDELINES Care should be taken that such an action does not expose the animal to excessive surgical or other invasive or painful procedures. The psychologist transferring animals should be assured that the proposed use by the recipient colleague has the approval of, or will be evaluated by, the appropriate institutional animal care and use committee and that humane treatment will be continued.
   2. It may sometimes be feasible to return wild-trapped animals to the field. This should be done only when there is reasonable assurance that such release will not detrimentally affect the fauna and environment of the area and when the ability of the animal to survive in nature is not impaired. Unless conservation efforts dictate otherwise, release should normally occur within the same area from which animals were originally trapped. Animals reared in the laboratory generally should not be released because, in most cases, they cannot survive or they may survive but disrupt the natural ecology.
B. When euthanasia appears to be the appropriate alternative, either as a requirement of the research, or because it constitutes the most humane form of disposition of an animal at the conclusion of the research:
   1. Euthanasia shall be accomplished in a humane manner, appropriate for the species, under anesthesia, or in such a way as to ensure immediate death, and in accordance with procedures approved by the institutional animal care and use committee.
   2. No animal shall be discarded until its death is verified.
   3. Disposal of euthanized animals should be accomplished in a manner that is in accord with all relevant legislation, consistent with health, environmental, and aesthetic concerns, and approved by the institutional animal care and use committee.

References

1. Emanuel EJ, Grady CC, Crouch RA, Lie RK, Miller FG, Wendler DD, editors. The Oxford textbook of clinical research ethics: Oxford University Press. New York, USA; 2008.
2. Smith T. Ethics in medical research: a handbook of good practice: Cambridge University Press. Cambridge, UK; 1999.
3. Resnik DB. What is ethics in research & why is it important. Natl Inst Environ Health Sci. 2011;1(10): 49–70.
4. Smith D. Five principles for research ethics. Monit Psychol. 2003;34(1):56.
5. Basic Information about Scientific Integrity, 2012, Feb. available at: https://www.epa.gov/osa/basic-information-about-scientific-integrity#definition.

6. Goodstein D. Scientific misconduct. Academe. 2002;88(1):28.
7. 10 Types of Scientific Misconduct, 2015, July 19. available at: https://www.enago.com/academy/10-types-of-scientific-misconduct/.
8. Shewan LG, Coats AJ. Ethics in the authorship and publishing of scientific articles. Int J Cardiol. 2010;144:1.
9. Abraham P. Duplicate and salami publications. J Postgrad Med. 2000;46(2):67.
10. Publication conduct available at: https://www.ndpublisher.in/publicationconduct.php.
11. Shamoo AE, Resnik DB. Responsible conduct of research: Oxford University Press. New York, USA; 2009.
12. Sugarman J, Sulmasy DP. Methods in medical ethics: Georgetown University Press. USA; 2010.
13. Beauchamp TL, Childress JF. Principles of biomedical ethics: Oxford University Press. New York, USA; 2001.
14. Ebbesen M, Andersen S, Pedersen BD. Further development of Beauchamp and Childress' theory based on empirical ethics. J Clin Res Bioeth. 2012;6:e001.
15. Avasthi A, Ghosh A, Sarkar S, Grover S. Ethics in medical research: general principles with special reference to psychiatry research. Indian J Psychiatry. 2013;55(1):86–91.
16. Code N. The Nuremberg Code. Trials of war criminals before the Nuremberg military tribunals under control council law, vol. 10. Washington, D.C: U.S. Government Printing Office; 1949. p. 181–2.
17. Carlson RV, Boyd KM, Webb DJ. The revision of the Declaration of Helsinki: past, present and future. Br J Clin Pharmacol. 2004;57(6):695–713.
18. Rice TW. The historical, ethical, and legal background of human-subjects research. Respir Care. 2008;53(10):1325–9.
19. Declaration Of Helsinki 1975. Available at: https://www.wma.net/what-we-do/medical-ethics/declaration-of-helsinki/doh-oct1975/.
20. Claudon M, Cosgrove D, Albrecht T, Bolondi L, Bosio M, Calliada F, Correas JM, Darge K, Dietrich C, D'onofrio M, Evans DH. Guidelines and good clinical practice recommendations for contrast enhanced ultrasound (CEUS)-update 2008. Ultraschall in der Medizin-European J Ultrasound. 2008;29(01):28–44.
21. Sims JM. A brief review of the Belmont report. Dimens Crit Care Nurs. 2010;29(4):173–4.
22. World Health Organization, Research ethics committees: basic concepts for capacity-building. 2009. available at: http://www.fercap-sidcer.org/publications/pdf/201202/FERCAP-19-WHO%20REC%20Basic%20Concepts.pdf.
23. LaFollette H, Shanks S. Brute Science. New York: Routledge; 1996.
24. American Psychological Association. Guidelines for ethical conduct in the care and use of animals. J Exp Anal Behav. 1986;45(2):127.
25. Moreno JD. Protectionism in research involving human subjects. In: National Bioethics Advisory Commission, editor. Commissioned papers and staff analysis. Bethesda: NBAC; 2001.
26. Manson NC, O'Neill O. Rethinking informed consent in bioethics: Cambridge University Press. New York, USA; 2007.
27. Dankar FK, Gergely M, Dankar S. Informed consent in biomedical research. Comput Struct Biotechnol J. 2019;17:463.
28. O'Neill O. Some limits of informed consent. J Med Ethics. 2003;29(1):4–7.
29. Morrow V, Richards M. The ethics of social research with children: an overview 1. Child Soc. 1996;10(2):90–105.
30. Knoppers BM, Chadwick R. Human genetic research: emerging trends in ethics. Nat Rev Genet. 2005;6(1):75.
31. Giesbertz NA, Bredenoord AL, Van Delden JJ. Consent procedures in pediatric biobanks. Eur J Hum Genet. 2015;23(9):1129.
32. Brall C, Maeckelberghe E, Porz R, Makhoul J, Schröder-Bäck P. Research ethics 2.0: new perspectives on norms, values, and integrity in genomic research in times of even scarcer resources. Public Health Genomics. 2017;20(1):27–35.

Further Reading

For a simple but comprehensive review of ethical theory, see Gensler HJ. Ethics: a contemporary introduction. Routledge; 2011.

For a booklet introducing core critical thinking concepts and principles as an empowering problem-solving framework, see Paul R, Elder L. The miniature guide to critical thinking concepts & tools. Rowman & Littlefield; 2019.

13

For a classic text in biomedical ethics, see Beauchamp TL, Childress JF. Principles of biomedical ethics. Oxford University Press, USA; 2001.

For discussions on applying evidence-based approach to research ethics see: Anderson EE, DuBois JM. The need for evidence-based research ethics: a review of the substance abuse literature. Drug Alcohol Depend. 2007;86(2–3):95–105; Kalichman M. Evidence-based research ethics. Am J Bioeth. 2009;9(6–7):85–7; Beagan B, McDonald M. Evidence-based practice of research ethics review. Health Law Rev. 2005;13(2–3):62–8; Anderson EE, DuBois JM. IRB decision-making with imperfect knowledge: a framework for evidence-based research ethics review. J Law Med Ethics. 2012;40(4):951–69.

For International principles and standards, see: Nuremberg Code: www.hhs.gov/ohrp/archive/nurcode.html; Council of Europe: Convention of Human Rights and Fundamental Freedoms http://conventions.coe.int/Treaty/en/Treaties/html/005.htm; Declaration of Helsinki: www.wma.net/en/30publications/10policies/b3/; International Compilation of Human Research Standards covering over 100 countries: www.hhs.gov/ohrp/international/intlcompilation/intlcompilation.html.

For International guidelines, see: Council for International Organizations of Medical Sciences (CIOMS) International Ethical Guidelines for Biomedical Research Involving Human Subjects: www.cioms.ch/publications/guidelines/guidelines_nov_2002_blurb.htm; UNESCO Universal Declaration on Bioethics and Human Rights: www.unesco.org/new/en/social-and-human-sciences/themes/bioethics/bioethics-and-human-rights/; The International Conference on Harmonisation's Guide to Good Clinical Practice (GCP): www.emea.europa.eu/pdfs/human/ich/013595en.pdf; World Health Organization Standards and Operational Guidance for Ethics Review of Health-Related Research with Human Participants: http://whqlibdoc.who.int/publications/2011/9789241502948_eng.pdf.

Correction to: Biomedical Product Development: Bench to Bedside, Learning Materials in Biosciences

Babak Arjmand (iD), *Moloud Payab* (iD), *and Parisa Goodarzi*

Correction to: B. Arjmand et al. (eds.), *Biomedical Product Development: Bench to Bedside,* **Learning Materials in Biosciences,** ► https://doi.org/10.1007/978-3-030-35626-2

The original version of the chapters 10 and 11 were inadvertently published with incorrect figures. Figure 1 of these chapters has now been updated.

The updated version of these chapters can be found at
https://doi.org/10.1007/978-3-030-35626-2_10
https://doi.org/10.1007/978-3-030-35626-2_11

Supplementary Information

© Springer Nature Switzerland AG 2020
B. Arjmand et al. (eds.), *Biomedical Product Development: Bench to Bedside*,
Learning Materials in Biosciences, https://doi.org/10.1007/978-3-030-35626-2

Index

Printed in the United States
By Bookmasters